Male and Female He Created Them

As a Complementary Partnership

Male and Female He Created Them

As a Complementary Partnership

by
Walter R. Dolen

Parisburg Publishing

Copyright 1976, 2014, 2025 by Walter R. Dolen
all rights reserved

First published in 1976 under the name:
Sex Makes the Difference: the Case Against Radical Women's Lib

2nd Ed. 2014: *Sex is the Difference:*
Case Against Radical Feminism is the Case for Real Feminism

2016: Updated printing:
Male & Female: Complementary Partnership

This 3rd Edition: Paperback
Male and Female He Created Them:
As a Complementary Partnership

ISBN: 978-1-61918-072-7

Digital copy at https://beone.ws/books.htm

Cover: in the public domain

1st printing November 2025

For this book and others by the author do a search on:
Amazon barnesandnoble.com Kindle B1

or do a search at the Library of Congress
or go to the author's web page: www.walterdolen.com

This book is a critique of radical feminism and sexual anarchy, as well as the writings, news articles, academic papers, podcasts and books they use and misuse for their dogma; all quotations cited are used with the protection of the fair use doctrine of the USA copyright law, and under the First Amendment to the Constitution of the USA.

Parisburg Publishing
Pennsylvania, USA

Dedication

This book is dedicated to all those in the future who will present, enact and enjoy a more realistic approach to sex relationships.

Contents

Contents	6
Documentation	10

Introduction — 11
Male and Female He Created Them — 11

1. Confusion and Misunderstanding — 15
Misunderstanding instead of Partnership — 15
What is Women's Lib? — 17
  Early Leaders — 17
  Some Results of Radical Feminism — 21
  Women Without Children — 25

2. Grievances And Claims of Feminism — 29
New Woman? — 29
Not a New Movement — 29
Do Feminists Speak For Most Women? — 32
Caricatured Wife — 32
Feminism, A Middle-Class Movement — 33
How Do Males Oppress Females? — 34
Marriage: The Enemy — 36
Freedom Through Work — or — Change Everything — 37
Separate-But-Equal Doctrine — 38
Working for Your Family v. Working for Others — 39
Meaningful Jobs — 41
Wage Differences — 41
Demand For Status Jobs — 42
One-Sided Equality of Women's Lib — 44
Affirmative Action — 45

3. Sex Roles — 49
Biology v. Culture — 49
  Sex Role Development — 49
Margaret Mead and Cultural Conditioning — 50
  Mead's Superficial Studies — 54
  Mead's Superficial Foreign Language Knowledge — 55
Other Points Against Mead's Work — 56
  Biased Work — 56
  Poorly Documented and Vague — 57
  Broken Cultures — 59
  "Matriarchal" Societies — 60
  Traditional Roles Still Prevailed — 61

Traditional Roles are in All Other Primitive Societies.	61
Persons in Reversed Roles were Maladjusted	61
Hormonal Levels not Studied	62
Experimental Societies' Disasters	62
Russia's Equality	63
The "Equality" in Israel's Kibbutz	65
Problems with Women in the Kibbutz	66
Traditional Family Tendencies in the Kibbutz	67
Kibbutz's Gender Equality	69
Sex Role Neutrality-at-Birth Theory	71
Beginning of the Transsexual Myth	71
Sex Role Assignment	73
Problems with Sex Role Neutrality-at-Birth Theory	74
Chapter 7 & 8 of Money's Book	78
Two Cases	78
Sex Role Development and Environmental Influences	82
Asocial & Prosocial Behavior	84
Functional Differences and Behavior	86
Women Have Babies; Men Do Not Have Babies	87
Real Reasons for Sex Roles	89
Science to Help?	89
Working Mothers with Children	91
What the Socialization Theory Cannot Explain	92
Conclusions on Sex Role Development	94

4. Real Sex Differences and Their Implications   98

What is Sex?	101
Sex Differences From Conception Onward	103
Sex and Hormones	104
Male & Female Hormones	105
Hormonal Interaction and Control	106
Brain Organizational Differences & Hormones	106
The Differences are Manifold	107
Intellectual Ability Differences — IQ Tests	110
Sex/Gender Differences in Verbal Ability, Number Ability, Spacial Ability Found in IQ Tests	111
Other Perceptual Differences	115
Erotic Perceptual Differences	118
Towards Sexual Maturity	121
Perceptual Distractibility	122
Maternal Behavior	122
Androgenic and Estrogenic Influence	124
Homeostasis	125
Hormones and Growth	126
Maturational Rate	126

Developmental Timetables	127
Homosexuality	129
Animal Research's Relevance	130
Physical Sex Differences	132
Genitalia	132
Form	132
Head Features	132
Breasts and Shoulders	133
Angle of Arms and Legs	133
Hips and Legs	133
Hands and Feet	133
Hair	133
Dimensions	133
Height	134
Weight	134
Strength	134
Vital Capacity and Muscular Tension	134
Vital Capacity	134
Muscular Tension	135
Nutrition	135
Biological Defects	135
Biology Limits and Causes Behavior	136
Absolute and Relative Limits	136
Biology is the Main Cause of Behavior	138
Biology is the Main Cause of Sex Behavior	139
Bio v. Culture Forces	139
Review	140

5. Motherhood and Maternal Deprivation — 144

Maternal Deprivation	145
How Does Maternal Deprivation Occur?	145
What is Wrong With Institutional Child Care?	146
Examples of Institutional Child Care	147
Effects of Maternal Deprivation on Children?	148
Harlow's Motherless Effect	149
Indifference of Unmothered Children Towards Motherhood	149
Childbearing and Childrearing	151
Wrong Motives for Childbearing	152
Nursing	153
Childrearing	155
How Should Children Be Reared?	156
Parenting by Homosexuals	158

6. Answer to Radical Feminism — 161

Feminism, Motherhood, Value, And Identity	161

To Feminists Men are The Value-Makers	162
Homosexuality Of Many Radical Feminists	163
Feminism As A Form Of Male Chauvinism	164
Answer To Feminism: Give Value Where Due	166
Sexes Need Sex-Differentiated Teaching	167
What Should Be Done	171

7. Sex Fallacies & Sexual Maturity — 173

Gossip	173
Superstition	173
Courage	173
Deception	173
Vanity	174
Words and the Degradation of Females	174
Achievement	174
Practical Sense	175
Equality	175
Men use Women	176
Perception, Self-Centered	176
Chauvinism	176
Sexual Maturity — The Goal	177

8. Summary of Evidence and Issues — 179

Conclusions on Sex Role Development	179
Socialization Theory Cannot Explain	180
Overlapping Rationale	181
Teaching and Sex Differences	183
50 Percent Hiring Demand Issue	184
Wage Difference Issue	184
Pregnancy and Work Leave Issue	184
Divorce, Child Support, and Alimony Issue	185
Abortion Issue	186
Sexual Freedom Issue	187
Dress Issue	187
Sexual Maturity	188

9. The Biblical Perspective — 189

What does the Bible say about males and females and sex?	189

BIBLIOGRAPHY — 191

Author — 202

Documentation

This book has both footnotes and endnotes. The *endnote* numbers are in brackets, such as [4]. Go to the end of each chapter to find the reference. For example, for endnote [4] in Chapter 1, you would go to the end of Chapter 1 and find:

[4] pp. 107-108 in **126**

From this you know that on pages 107 to 108, in the document numbered '126' in the Bibliography, you will find the appropriate source.

The source information for *footnote* numbers that are superscripted (as the 1 in "footnote[1]") are found at the bottom of each page.

In future Kindle, e-book or PDF versions there will be hot links to some of the footnotes/endnotes and sources.

Introduction

Male and Female He Created Them

In the Holy Bible, Genesis 5:2 it says:
"Male and Female He Created them"

If God[1] created them male and female, why all the commotion about transsexuals in Western politics, newspapers and colleges. This is a good question to ask, if you are a Christian or are not a sexual anarchists. Is there any real proof that there are more than two biological sexes (or genders)? There is confusion here because of the words "gender" and "sex." What is the difference? Decades before the 1960s and early1970s the word "gender" was not used the way it has been used recently. It was only used in grammar as pronouns to depict someone who was either male ("he") or female ("she") or neither ("it"). The word "gender" was also used to classify things into certain classes. But that's it. If you look at any dictionary printed before 1960s under the word "gender," there weren't hundreds of sexual *genders* as some recently try to portray. When the sexual-anarchists speak of 100s of "genders," they are really speaking about the 100s of sexual fantasies or fetishes as describes in Nancy Friday's books.[2] The reason gender was sexualized was because of a few individuals in the 1960s and 1970s pushed their own mistaken belief in the — "sex role neutrality-at-birth theory." We successfully destroy this myth of multiply genders/sexes in this book. We also manifest the evidence that indeed there are only males and females and that basically biology dictates the sex roles and behavior of both. Biology of males and females are complementary, a beautiful partnership of roles and biology made by God that also depicts and representative of the image of God.[3] The vast majority of this book uses scientific studies as well as the words of the proponents of feminists, sexual anarchists, and their subjects. This work is not a Biblical study, except for the last chapter.

Male and Female is a book that points to the natural complementary partnership of the sexes by manifesting the real

---

[1] The *All-Powerful-Being* who created the universe. See *God Paper* for definition.

[2] https://www.amazon.com/stores/Nancy-Friday/author/B000APBBJ4?ccs_id=bd543d74-b42a

[3] see the *God Papers* book and the "Image of God" paper. https://beone.ws/books.htm

sexual/gender differences between the sexes. It is the case against the naivete of radical feminism and their proponents. The differences between the sexes are not celebrated by them. The feminists believe that sex/gender differences are caused almost exclusively by the social environment. This book gives the evidence against these mistaken beliefs on socialization while embracing the real women — the ones who are the backbone of life on this planet. They are the ones not given enough credit for their role by the feminists. This book points to the complementary roles of the sexes by manifesting that the sexes are different in some ways, but these differences are also complementary, thus a natural partnership exists in their walk through life.

This book examines the nurture/socialization ideas and the nature/biological ideas pertaining to sex/gender differences. Although socialization may slightly magnify or diminish the differences, it is the underlying and persistent biology that is the source of real sex/gender differences, not socialization. Despite the obvious difference — women bear children (men don't) — radical feminists ignore and work against the obvious and call for *total* "equality." The feminists say that in the past, superficial understanding of these differences was used against women to "hold" them down. In their dialogue being *held down* means to keep them at home out of the workplace, as if being a mother was not all that important. They say women were subjected to the idiocy of Sigmund Freud and other psychoanalysts and subjected to naive interpretations of Darwin's theory, which they were, but feminists get this critique right for the wrong reason. The feminists say women were subjected to clitoridectomy and removal of the uterus and ovaries to fix her sexual and psychological problems, which is correct, but again for the wrong reason. The real reason for these atrocities was the general lack of knowledge of real sex differences by the so-called scientific world — run by men — and the prevailing male chauvinistic mindset. Tragically, women's role of motherhood was denigrated by both the male chauvinists and the feminists. From an 1856 book we get an idea of the negative thinking against women back then:

> Woman is now bound hand and foot by custom and law. She is only a thing. She is not a conscious independent personality. She is not recognized as a self-directing, responsible agent. She plays a secondary part. She is shut out from all the higher aims and opportunities of life. Into no college is she permitted to enter.... Every profession but the teacher's is barred against her, and in that her services are

considered not half at par. She can not get more than half-pay for her labor.[4]

This thinking is also prevalent today in some religions, cultures and nations, especially among those who are ignorant of the complementary sex differences, which include the feminists and the male chauvinists.[5]

Feminists materialized to fight the negative naive ideas and biases pertaining to females, and to push sexual anarchy. The problem though is that *radical* feminism has made inroads in our laws and schools while ignoring real biological differences. This movement set up unrealistic expectations that have caused more harm than good. We believe acknowledgement and acceptance of the real differences and needs of each sex, is the answer.

The early foundation of this work was an intense study of sex differences in the early 1970s in the libraries of the University of California at Berkeley, Stanford University in California and the San Jose State University. Why was I studying this subject? Because several of my girl friends were espousing the ideology of the women's lib movement in the late 1960s and early 1970s. Their main premise was that there were no real differences between the sexes, only differences in socialization. To see if this was true, I went and looked for books on the subject. I went from library to library, book to book, from article to article, from study to study. I spent two years collecting evidence pro and con on the subject of sex differences. From this information, I wrote a book on the subject called *Sex Makes the Difference in* 1976. Since then, the scientific studies have continued by academia (howbeit providing more evidence against radical feminists' myths) with more information about the effects of hormones on males and females, and how males and females use their brains differently as seen through fMRI scans.[6] After my first study I continued to study and collect material through the years. In 2014 I incorporated some of it in the second edition and changed the

---

[4] G.S. Weaver, *Aims and Aids for Girls and Young Women of the Various Duties of Life...*, Fowler and Wells, Publishers, London & Boston, 1856. p. 15. The author was describing women's status in 1856, but his book was written to encourage his readers (women) to "be strong in mind and purpose...the time is not far distant when she must be educated as well as man...."

[5] Read the whole book and see especially "Feminism as a Form of Male Chauvinism" [Chapter 6] and "Chauvinism" [Chapter 7].

[6] Functional magnetic resonance imaging = fMRI: is a functional neuroimaging procedure using MRI technology that measures brain activity by detecting associated changes in blood flow.

title to *Sex is the Difference.* In 2016, I again changed the name of the book to *Male & Female: Complementary Partnership* to better describe the real contents of the book. In 2023 I updated the footnotes. Now in 2025 I have again updated the footnotes, added more information and added a new chapter to the book. To write a comprehensive report and summary of all the studies on biological sex differences would take several large volumes. Someone should do it, but even if they did, I doubt if many would read it. I have endeavored to put as much information in this approximately 200+ page book as possible, so perhaps some will read it and learn of the possibilities for harmony and understanding between the sexes.

— Walter
October 2025

1. Confusion and Misunderstanding

Misunderstanding instead of Partnership

The truth is that one half of the world has difficulty understanding the other half. Males misperceive females; females misperceive males. What do women want? Men only want sex. Women nag too much. Men are not sensitive enough. Women are too emotional. Men don't show their emotions. He doesn't appreciate me. She shows me no respect. Women cry too much. Men are taught not to cry. And so on.

In this book we will look at biological sex differences, sex/gender[7] roles, radical women's lib, men's views and treatment of women, women's views and treatment of men and other aspects of the interaction of males and females. Why examine these subjects? Because — there is potential for better harmony and understanding between the sexes that would lead to a complementary partnership. Real male and female differences are complementary,[8] not adversary. When both sexes understand this, it should ultimately lead to better harmony between the sexes with better family life and healthier children.

Something happened in the late 1960s and early 1970s. The birth-control pill, women's lib, legalized abortion, more women going into the workforce and the sexual revolution took place, among others things. We are going to examine what radical "feminism" and the shifting attitudes of both men and women had to do with the changes that took place. Of course, the world has never been a utopia. Life was hard for all generations in different ways. But with the industrial and electronic revolution, life should have become much better. Instead of a one-wage family working only 32 hours per week *or less* as promised by the elite in the media of the 50s & 60s,[9] now, most families have been turned into two-income families[10] —

---

[7] "Sex" and "gender" have mixed meaning in literature, even in the professional journals. The use of the word gender instead of sex has been attributed to the influence of feminist studies and is sometimes used as a euphemism for sex in non-feminist writings.

[8] *Complementary* — "completing something else or making it better; serving as a complement —used of two things when each adds something to the other or helps to make the other better; going together well: working well together."

[9] "16-Hour Work Week by Year 2020," *Gastonia Gazette*, Nov 26, 1967; etc.

[10] Only about 28% are one-income families with children under 18, "America's Families and Living Arrangements: 2013," Table FG2, *US Census Bureau*

both together working 60-80 plus hours a week, with less leisure. What happened? For one thing, when women went into the workforce[11] (out of necessity, out of social pressure[12]), real wages dropped,[13] making it almost impossible for stay-at-home mothers to stay at home and make a comfortable home for their families:

> When we consider all working-age men, including those who are not working, the real earnings of the median male have actually declined by 19% since 1970. This means that the median man in 2010 earned as much as the median man did in 1964 — nearly a half century ago. Men with less education face an even bleaker picture; earnings for the median man with a high school diploma and no further schooling fell by 41% from 1970 to 2010.[14]

Also when women's wages in the USA were allowed to be counted to qualify to buy family homes,[15] the prices of homes increased dramatically in the 1970s,[16] which made it almost impossible for single income families to buy houses, thus forcing more women into the marketplace. And instead of computers and automation *et al* helping to add leisure to the families, the family unit is now under great stress with less leisure. Yes, not all of these problems were caused directly by radical women's lib ideas, but a lot were, at least indirectly.

---

[11] US Bureau of Labor Statistics: from 22 million in 1970 to 31 million in 1980, about a 50% increase in the 1970s alone

[12] by Women's Lib and by men who wanted income from their wives to buy more things

[13] "What Happened to the Wage and Productivity Link", *Outside the Beltway*, July 22, 2012

[14] "The uncomfortable Truth About American Wages," *New York Times*, Oct 22, 2012

[15] Equal Credit Opportunity Act of 1974, 15 U.S. Code § 1691: This act made it illegal for financial institutions to discriminate against female applicants just because they were married and were in the childbearing age. Before this law banks only allowed a portion of married women's wages, who were of childbearing years, to count for qualifying for home loans because it was assumed women would eventually quit their jobs to help raise the children. http://www.law.cornell.edu/uscode/text/15/1691

[16] in the 1970s family income doubled (more wives went to work), while home prices tripled (quadrupled in California). 1970: http://www.1970sflashback.com/1970/Economy.asp / 1980: http://www.1980sflashback.com/1980/Economy.asp ; Calif: http://www.realestateabc.com/graphs/calmedian.htm

1. Confusion and Misunderstanding

What is Women's Lib?

Radical Feminism v. Real Womanism. According to women libs of the 1960-70s, women were being treated unfairly in societies throughout history: denied education; treated as second-class citizens; given lower status/worth by men *and* women themselves; etc. Feminism in the 1960s grew to counter these apparent unfair conditions. There were many books and articles available on the women's lib movement in the 1960s and early 1970s. These books and articles became the sounding board for some of the radical aspects of women's lib as well as for examining real unfair discrimination against women. In the more moderate written works, women's lib may have come across as a logical movement based on a solid foundation of facts which could bring equality to women. But to others, such as housewives, the movement was "not a movement calling for equal opportunity, equal pay, equal status for women's role in life, in fact as well as in law; instead it attacks the very nature of women, and in the guise of liberation, seeks to enslave her."[1]

Radical Lib. What is the radical women's lib or radical feminism? As in its name, it is radical. The radical feminists were using the false theory of *sex-role-malleability* in their attempt to forcefully change the world to fit their dogma. Their ultimate goal was/is to rid the world of all sex roles and sex differences because they believed most sex differences were culturally determined. We examine this in some detail in Chapters 3 & 4.

There is an important difference between radical feminism and the real role of women. The key word here is *radical*, not feminism, yet we will use "feminism" often to mean "radical feminism" for that is how it is mostly understood today.

Early Leaders

To examine the nature of the movement let's look at what some of the early leaders of this movement said. Of course, like any movement or ideology, some of it made sense; some of it was radical. We will directly quote from such early feminists as Elizabeth Janeway, Mary Mothersill, Elizabeth Cady Stanton, Aileen S. Kraditor, Annette Grant, and Simone de Beauvoir:

- "Complaints about the movement are many: it hasn't defined its issues clearly, it differs within itself, its goals are either utopian or minuscule and, above all, it traffics in emotion instead of logic. Yet it persists!...

    "If the women's movement cannot be easily contained within any set definition nor held to any stated program, perhaps that is because it is larger and more novel than it has been thought to be. I believe that to be true. The movement seems to me to be a response to profound and irreversible historical forces involving economic, technological, and scientific shifts in our society. No wonder it hasn't yet found itself a satisfactory name or a coherent ideology!"[2] (Janeway)

- "Whether or not the claim that women are and have always been oppressed by men, can be made good — something that would require a more satisfactory analysis of "oppression" than is currently available — it is clear that in particular places at particular times, for example, here and now, women are treated unfairly. Once this fact is recognized, wherein lies the need of theory?"[3] (Mothersill)

- "Many times and oft it has been asked of us, with unaffected seriousness, 'What do you women want? What are you aiming at?'... We ask for all that you have asked for yourselves in the progress of your development...."[4] (Stanton)

- "According to Stanton, what some women — the enlightened ones — want, and what all women deserve is equality of status and opportunity in political, social and domestic contexts. To critics who objected to the vagueness and generality of her demand, she replied with a list of specific legislative reforms, including woman suffrage but extending to education, property rights, employment practices and divorce law."[5] (Mothersill)

- "Feminism is customarily thought of as the theory that women should have political, economic, and social rights equal to those of men.... Clearly, the history of American feminism implies far more that the practical application of the theory stated above — that women should have rights equal to men's. What the feminists have wanted has added up to something more fundamental than any specific set of rights or the sum total of all the rights that men have had.

    "The fundamental something can perhaps be designated by the term 'autonomy.' Whether a feminist's demand has been for all the rights men have had, or for some but not all of the rights men have had, or for certain rights that men have not had, the grievance behind the demand has always seemed to be that women have been regarded not as people but as female relatives of people. And the feminists' desire has, consequently, been for women to be recognized, in the economic, political, and/or social realms, as individuals in their own right."[6] (Kraditor)

- "The purpose of NOW is to take action to bring women into full participation in the mainstream of American society now, exercising all the privileges and responsibilities thereof in truly equal partnership with men."[7] (The 1966 Statement of Purpose of the National Organization for Women)
- "What is Women's Liberation all about anyway?...It's about women's desire — demand — to be free to be women — free to define themselves instead of being 'set pieces' in our society."[8] (Grant)
- "What is certain is that hitherto woman's possibilities have been suppressed and lost to humanity, and that it is high time she be permitted to take her chances in her own interest and in the interest of all."[9] (de Beauvoir)

From these quotes and the hundreds of others we read in our research, we can see that these women liberators wanted freedom. They wanted freedom from alleged male oppression; they wanted freedom to be just like men. According to them, the male oppression comes in all forms from the way men perceive women to the way they make love to them. The freedom to be like men comes in all forms from the possession of similar jobs to dressing like men.

Homosexuals. There is also the factor of homosexuality in the women's lib leadership. Notice:

> A vital relationship between lesbians and Women's Liberation is in their mutual interest in a time of changing relationships. Lesbians are the women who potentially can demonstrate life outside the male power structure that dominates marriage as well as every other aspect of our culture. Thus, the lesbian movement is not only related to Women's Liberation, it is at the very heart of it.[17]

---

[17] Sidney Abbott and Barbara Love, "Is Women's Liberation a Lesbian Plot?," p.620, *Woman in Sexist Society*, 1972

Even though homosexuals only make up about 1.6 to 3% of the population,[18] they played a large part in the movement, especially in the books, articles and speeches. Some of the real energy and myths behind the radical aspects of the movement were from the female-homosexuals, or from those with male-like traits (possibly caused by androgen-like substance introduced during their formation in the womb[19]), or from just plain rebels.

Against Family. One big aspect of radical women lib was their biased attitude pertaining to marriage and family:

- Marriage has existed for the benefit of men; and has been a legally sanctioned method of control over women.... We must work to destroy it. The end of the institution of marriage is a necessary condition for the liberation of women. Therefore it is important for us to encourage women to leave their husbands and not to live individually with men... All of history must be re-written in terms of oppression of women. We must go back to ancient female religions like witchcraft.[20]

- The traditional family, with all its supposed attributes, enslaved woman; it reduced her to a breeder and caretaker of children, a servant to her spouse, a cleaning lady, and at times a victim of the labor market as well.[21]

In fact, marriage is anything but detrimental for women. Most women like being married, proudly wear their wedding ring and one of the most important days in their life is their wedding day. Marriage, before no-fault divorce, actually protected women and gave them an atmosphere to raise their children in some kind of security.[22] Instead of raising their offspring alone, like most females of the animal kingdom, marriage gives each woman a mate to help and be loved by. This was and is especially important for humans since their children take a much longer time to get to the point where

---

[18] "Sexual Orientation and Health Among U.S. Adults," CDC — http://www.cdc.gov/nchs/data/nhsr/nhsr077.pdf ;

[19] see Chapters 3 and 4

[20] Declaration of Feminism, November 1971

[21] Sol Gordon, "The Egalitarian Family is Alive and Well," *The Humanist*, May/June 1975, p. 18

[22] Although some of the men who had to pay alimony were unfairly treated

they become independent of their parents. This was always understood and thus, in the original Universal Declaration of Human Rights of the United Nations (1948) Article 16: "The Family is the natural and fundamental group unit of society and is entitled to protection by society and the state."[23]

Today, the same words still exist in the Declaration. But because marriage is undermined and deemphasized by more than some of the elites, both women and children are being hurt, as well as the well-being of the nations. As we examine the radical women's lib movement you will see much more of their extreme anti-family bias.

From the radical aspects of the movement came the greater disharmony and conflict in the relationship between the sexes and in family life.

Some Results of Radical Feminism

More Abortions. Along with the women's lib movement came abortion. Legalized abortions dramatically increased from 5,000 in 1963, 18,000 in 1968, 236,000 in 1970, 586,000 in 1972, 616,000 in 1973 to about 1,600,000 during the late 1980s and down to about 1,000,000 in 2011.[24,25] Reporting of abortions is not mandatory so statistics may vary from reality: thus the number of "reported" abortions to the Centers for Disease Control and Prevention (CDC) is slightly lower than estimates by the Guttmacher Institute. Although no one really knows, illegal abortions before the 1973's Supreme Court *Roe v. Wade* decision have been estimated from tens of thousands to much higher figures.[26] Almost four times as many black women as non-Hispanic white women had abortions in 2010 (CDC). Half of pregnancies among American women are unintended, and four in 10 of these are terminated by abortion. It is estimated that one-third of all women in the US will have an abortion by the age of 45. We can't blame all the increase in abortions on feminism, but it

---

[23] Universal Declaration of Human Rights, Dec 10, 1948, by the general Assembly of the UN, Article 16: https://www.un.org/en/about-us/universal-declaration-of-human-rights

[24] Table 2, "Morbidity and Mortality Weekly Report," CDC, Nov 26, 2004

[25] "Induced Abortion in the United States," Guttmacher Institute, July 2014 — http://www.guttmacher.org/pubs/fb_induced_abortion.html

[26] "Lessons from Before Roe: Will Past be Prologue?," *The Guttmacher Report on Public Policy*, March 2003. — http://www.guttmacher.org/pubs/tgr/06/1/gr060108.html

had a great deal to do with this increase since, as we will see, radical feminism emphasized working outside the home and freedom from the "bondage" of marriage and motherhood. Also, single men in relationships pressured their girlfriends to get abortions. While previously these same men were forced into marrying their impregnated girlfriends, which in many cases led to early divorces and unhappy marriages. Thus, abortion aided the new lifestyle for both men and the "new" women.

Marriages and Divorces. With the new freedoms came fewer marriages and more divorces. In 1960 there were four marriages each year for each divorce. The 1960 figures were bad enough when compared to 1940 when there were six marriages each year for each divorce. But today, according to the statistics, there are only about two marriages each year for each divorce. Does this mean that 50% of marriages end in divorce? No. According to a U.S. Census Bureau 2011 study of the men and women aged 40-49 only about 30% were ever divorced (60-69 about 35% were divorced; 70 and over only about 22% were ever divorced).[27] The reason for this apparent contradiction between the 50% divorce rate and the 22% to 35% rates is the confusion over the various ways to view statistics. Despite the confusion, since the 1960s divorces have increased, especially for the so-called baby boomers. Also, children of divorced parents are four times more likely to become divorced themselves.[28]

Fatherless Children. And with the new freedoms of feminism came "free" sex and its consequence — fatherless children. In 1960 87% of children lived with two parents; in 1990 only about 72% of children lived with two parents. Today, over 30% of all children live with only one of their parents. Over 55% of black children are brought up without fathers.

What are the damages caused directly or indirectly by children having no fathers in their life:
- 85% of children who exhibit behavioral disorders come from fatherless homes. [Center for Disease Control]
- 85% of youths in prisons grew up in a fatherless home. [Fulton County Georgia jail populations, Texas Department of Corrections, 1992]

---

[27] Table 6, p. 16 — "Numbers, Timing, and Duration of Marriages and divorces: 2009," Issued May, 2011, Source: U.S. Census Bureau, Survey of Income and Program Participation (SIPP).

[28] Shapiro, Andrew LL., We're Number One, Vintage Books, New York, 1992, p. 35.

1. CONFUSION AND MISUNDERSTANDING

- 90% of homeless and runaway children are from fatherless homes. [US D.H.H.S., Bureau of the Census]
- 63% of youth suicides are from fatherless homes (US Dept. Of Health/Census) – 5 times the average
- 80% of rapists with anger problems come from fatherless homes –14 times the average. (Justice & Behavior, Vol 14, p. 403-26)
- 71% of pregnant teenagers lack a father. [U.S. Department of Health and Human Services press release, Friday, March 26, 1999]
- 90% of adolescent-repeat arsonists live with only their mother. [Wray Herbert, "Dousing the Kindlers," Psychology Today, January, 1985, p. 28]
- 71% of high school dropouts come from fatherless homes. [National Principals Association Report on the State of High Schools]
- 75% of adolescent patients in chemical abuse centers come from fatherless homes. [Rainbows f for all God's Children]
- 70% of juveniles in state-operated institutions have fatherless homes. [US Department of Justice, Special Report, Sept. 1988]
- Fatherless boys and girls are: twice as likely to drop out of high school; twice as likely to end up in jail; four times more likely to need help for emotional or behavioral problems. [US D.H.H.S. news release, March 26, 1999][29]

Never Married and Children. In 1960, 8.7% of females and 16% of males 29 years-of-age had yet to marry. In 1990 23.5% of females and 36% of males 29 years-of-age had never married. In 2010, 27% of the females 30-35 had yet to marry; males were at 36%. In 2016 about 36% of females 30-34 were unmarried while 45% of males were likewise unmarried.[30] In 1940, there were only about 110,000 births to unmarried women, while in 1960 there were approximately 250,000 such births (about 5%). In 2011, about 41% of the births

---

[29] US Dept of Health & Human Services, Bureau of Census — info taken from: http://www.menstuff.org/issues/byissue/fatherless.html#book; http://www.menstuff.org/books/byissue/divorce-general.html#kc1 ; etc.

[30] https://www.census.gov/programs-surveys/acs.html

were to unmarried women or about 1,6000,000: 29% whites; 53% Hispanics; 72% blacks. In 2021 about 40% of births were to unmarried women.[31] In about 58% of these the women were in a cohabiting union.[32] Today in France and England the figure is about the same, but more of them eventually marry so a higher percentage of them are reared with married parents than in the United States.

Mothers Working Outside The Home. With the new emphasis on working outside the home came more women working outside the home. In 1870 women were 15% of the workforce.[33] In 1967, women made up 30% of the full-time labor force and 33% of all workers (full & part time).[34] In 2009 women made up approximately 44-46% of our total working force outside the home.[25] In 2011, 58% of all women over 16 years of age work verses 64% of all men over 16 years.[35]

In 1948, approximately 10.7% of married women with children **under six** years of age worked, in 1955 16%, in 1960 18.6%, in 1970 33%, but in 2013 the figure is 64%.[36] In 2010, about 71% of all mothers with children **18** and under work outside the home.[37]

Children in Poverty. In families headed by single women in 2012 about 56% were classified in poverty by governmental statistics; in 1960 it was 24% classified in poverty.[38]

---

[31] https://www.cdc.gov/nchs/data/nvsr/nvsr72/nvsr72-01.pdf

[32] Child Trends Databank. (2013). *Births to unmarried women.* Available at: http://www.childtrends.org/?indicators=births-to-unmarried-women ; and http://www.cdc.gov/nchs/data/databriefs/db18.pdf

[33] 1870 US Census

[34] Table A-4, "Income, Poverty, and Health Insurance Coverage in the United States: 2009— http://www.census.gov/prod/2010pubs/p60-238.pdf

[35] Table 1, "Women in the Labor Force: A Databook," US Bureau of Labor Statistics, Feb 2013

[36] US Department of Labor, Bureau of Labor Statistics. www.data.bls.gov; extrapolated from http://www.infoplease.com/ipa/A0104670.html

[37] Bureau of Labor Statistics — http://www.bls.gov/news.release/famee.nr0.htm see also http://www.infoplease.com/ipa/A0104670.html

[38] Table 10, "Related Children in Female Householder Families, by Poverty Status: 1959 to 2012," US Census Bureau — http://www.census.gov/hhes/www/poverty/data/historical/people.html

1. Confusion and Misunderstanding

Women Without Children

Today millions of career women will never have children, never get to teach their children what they know, never pass on their intellectual qualities and abilities, never get to love and raise their own children, never get to see their grandchildren, never get to leave their genes to others. Many are unhappy about it, as they should be. It is their one big regret:

> Many mid-career women blame the movement for not knowing and for emphasizing the wrong issues. The ERA and lesbian rights.... The bitterest complaints come from the growing ranks of women who have reached 40 and find themselves childless, having put their careers first. Is it fair that 90% of male executives 40 and under are fathers but only 35% of their female counterparts have children? 'Our generation was the human sacrifice,' says Elizabeth Mehren, 42, a feature writer for the Los Angeles Times. 'We believed the rhetoric. We could control our biological destiny.... Nonprofessional women, poor women, minority women feel their needs and values have been largely ignored by the organized women's movement, which grew out of white, middle-class women's discontent.... Ask a woman under the age of 30 if she is a feminist, and chances are she will shoot back a decisive, and perhaps even a derisive, no.[39]

Notice what Barbara Walters said on the "Piers Morgan Live" show on December 17, 2013, as reported by the *Deseret News:*[40]

> Asking Walters to reflect on a career that has included trailblazing as the first female anchor on an evening news program, three Emmy Awards and interviews with every U.S. president since Richard Nixon, Piers posed this question:
>
> "Final question. It's kind of not a best or worst. It's more like, you've had such an extraordinary life and career, and it's continuing on into your retirement, and I'm sure it will carry on after that. If you could relive one moment in your life, the moment that brought you the greatest satisfaction, thrill,

---

[39] pp. 82, 81, *Time*, Dec. 4, 1989; in 2014 some women aren't having children because more husbands don't want them, which is another problem that needs attention.

[40] http://www.deseretnews.com/article/865592722/Not-making-time-for-motherhood-The-one-thing-Barbara-Walters-regrets.html; my emphasis

sadness perhaps. I mean, what has been, you think, the moment?"

Without waiting for Morgan to complete the question, Walters asked if she could instead express a regret.

"I regret not having more children. I would have loved to have had a bigger family," Walters said.

Barbara never had any children of her own, just one adopted child. She missed out on motherhood because of her career as did millions of others who followed in the path of radical feminism:

> The latest numbers suggest that an amazingly high percentage of women today—18.8 percent—complete their childbearing years having had no children. Another 18.5 percent of women finish having had only one child. Together, that's nearly **40 percent of Americans who go their entire lives having either one child or no children at all**.[41]

Are you a Feminist? In 1989 Time/CNN surveyed the question, "Do you consider yourself a feminist?" In response, 33% of the women said yes while 58% said no, and 76% of the women said they paid little or no attention to the women's movement.[42] In 1992, another Time/CNN survey asked the same question, "Do you consider yourself a feminist?" 29% of the women said yes while 63% said no.[43] Today most women are leery of calling themselves a "feminist" for different reasons, such as its anti-male and radical connotations. In April 16, 2013, the *Huffington Post* did a survey[44] of 1000 people — just 23% of the women considered themselves feminists.

Yes, there were and still are problems and misunderstandings between the sexes. Instead of more harmony, radical feminism led to more strife between the sexes. Because the radicals went too far, lives have become more complex for each sex as well as their children.

---

[41] my emphasis; http://www.weeklystandard.com/blogs/rise-childless-americans_654984.html; Statistics from the CDC and http://www.census.gov/hhes/fertility/data/cps/2010.html

[42] p. 89, *Time*, Dec 4, 1989

[43] *Time*, March 9, 1992, p. 54

[44] Emily Swanson, "Poll: Few Identify As Feminists, But Most Believe In Equality of Sexes," http://www.huffingtonpost.com/2013/04/16/feminism-poll_n_3094917.html

Stress and tension have increased according to the Department of Labor:

> Stress and the tension between work and family are increasing. Major changes in American families—and the lack of corresponding changes in many workplace policies and practices—are the causes.[45]

Increase in mental problems. Unrealistic expectations were introduced since the 1960s, and when people fail to reach these goals their happiness is reduced, and mental problems increase. Psychological problems have increased for men, women and children: 26% of women, 15% of men, 7% of boys and 5% of girls were using mental health medications in 2010.[46] Nearly 23% of women between 40 and 59 took antidepressants in 2005-08,[47] which was a 400% increase from the previous decade. The stress of women taking on careers, while still trying to nurture their children (both, together, an 80 hour plus job), is one of the causes. A Grok search on October 1, 2025 showed similar results except females had huge increases in antidepressant prescriptions after the pandemic as much as 130% in girls ages 12-17. In contrast, boys only increased 7%. Females (18-25) antidepressant prescriptions increased 57% while males rates showed minimal change after the pandemic.

In order to further ascertain the true significance and meaning of the radical women's lib movement, we need to examine some of the details of the grievances and claims of radical feminism that came up in the 1960s and 1970s.

References for Chapter 1

[1: ch 1] p. 4 in **177** of the Bibliography list
[2: ch 1] p. 91 in **88**

---

[45] "Work and Family," US Department of Labor

[46] "America's State of Mind," A Report by Medco — http://www.toxicpsychiatry.com/storage/Psych%20Drug%20Us%20Epidemic%20Medco%20rpt%20Nov%202011.pdf

[47] *Time*, Oct 20, 2011, from a CDC study, a 400% increase from decade before

[3: ch 1] pp. 113-114 in **126**
[4: ch 1] pp. 107-108 in **126**
[5: ch 1] p. 109 in **126**
[6: ch 1] pp. 7-8 in **95**
[7: ch 1] p. 363 in **130**
[8: ch 1] p. 16 in **70**
[9: ch 1] p. 679 in **42**

2. Grievances And Claims of Feminism

New Woman?

Early in the contemporary women's lib movement we heard about the emerging NEW woman. We saw pictures and stories of girls breaking into little league, or women being placed in some management positions, or heading some top government commission. We heard the so-called enlightened educators assure us that through proper education and socialization women too could become just like men. And through this, they imply, women would gain equality. We heard about the future, and its test-tube babies that would allow women to be emancipated from "maternal bondage." We read about women's liberation as some NEW movement, with NEW ideas, for the NEW woman. But the feminist movement even in the late 1960s and early 1970s was not new.

Not a New Movement

By looking at the history we can see this movement existed thousands of years ago. You can see it in such sources as the Bible, in the evolution of the Roman Empire's codes of law, and through various papers and letters of radical feminists. Such books as *Up From the Pedestal* [1] and *Feminism: The Essential Historical Writings* [2] have various feminist themes in them.

One of the earliest recorded calls for women's rights took place sometime around 1450 B.C. The daughters of Zelophehad of Israel came to Moses and spoke as follows:

> "Our father died in the wilderness.... He died... and left no sons. Is it right that, because he had no son, our father's name should disappear from his family? Give us our property on the same footing as our father's brothers."[3]

This movement by the daughters of Zelophehad to obtain the property of their father on the same footing as the brothers was a successful one. This inheritance problem was cleared up because the daughters of Zelophehad got together and demanded their rights.

In, *The Second Sex*, Simone de Beauvoir traces some of the liberalization of the Roman Empire's laws concerning women:

> Legally [the women were] more enslaved than the Greek, [yet] the woman of Rome was in practice much more deeply integrated in society.... She directed the work of the slaves; she guided the education of the children....She shared the labor and cares of her husband, she was regarded as co-owner of his property....The tie that bound her to him was so sacred that in five centuries there was not a single divorce.... On the street men gave them the right of way.... Little by little the legal status of the Roman woman was brought into agreement with her actual condition....The domestic tribunal disappeared before the public courts. And woman gained increasingly important rights. Four authorities had at first limited her freedom: the father and the husband had control of her person, the guardian and the *manus* of her property.... [de Beauvoir then continues on and describes the liberalization of the laws until finally —]
>
> ...She could inherit, she had equal rights with the father in regard to the children, she could testify. Thanks to the institution of the dowry, she escaped conjugal oppression, she could divorce and remarry at will....The Roman women *demonstrated*: they swarmed tumultuously through the city, they besieged the courts, they fomented plots, they raised objections, stirred up civil strife....
>
> When the collapse of the family made the ancient virtues of private life useless and outdated, there was no longer any established morality for woman....there were many women who refused maternity and who helped to raise the divorce rate....
>
> ...They meddled in politics, plunged into the files of legal papers, disputed with grammarians and rhetoricians, went in passionately for hunting, chariot racing, fencing, and wrestling. They were rivals of the men, especially in their taste for amusement and in their vices....[4]

2. Grievances and Claims

This description by de Beauvoir sounds like the radical feminists of today — they were the radical feminists of their day. [I can't guarantee that de Beauvoir's history is totally correct; remember she was a radical feminist, but it is interesting.]

In feministic historical writings, arguments for liberalizing of dress customs, granting women the vote, giving them higher education, and giving them equal rights to economic independence are abundant. Arguments comparing women with slaves, and using various scientific discoveries to reason for further liberalization are also in feminist writings of the 19th century.

Alexis de Tocqueville. Even in Alexis de Tocqueville's writings of the 1830s we see signs that the movement was alive in Europe at that time:

> "There are people in Europe who, confounding together the different characteristics of the sexes, would make man and woman into beings not only equal but alike. They would give to both the same functions, impose on both the same duties, and grant to both the same rights; they would mix them in all things — their occupations, their pleasures, their business."[5]

In fact, at the very beginning of our country some women were calling for equal rights, as well they should have. Abigail, John Adam's wife, wrote the following to her husband in April of 1776:

> "In the new code of laws which I suppose it will be necessary for you to make, I desire you would remember the ladies and be more generous and favorable to them than your ancestors. Do not put such unlimited power in the hands of husbands. Remember, all men would be tyrants if they could. If particular care and attention is not paid to the ladies, we are determined to foment a rebellion, and will not hold ourselves bound by any laws in which we have no voice of representation."[6]

Again, I am not saying women didn't and don't have legitimate grievances; it is the illegitimate grievances and the adversariality of the radical feminist movement that needs adjustment so that a more complementary and harmonious society may emerge.

Do Feminists Speak For Most Women?

Radical feminists think and act like they speak for all women. They think of themselves as the "enlightened ones" who are fighting for "what all women deserve."[7] Yet they admitted at the same time that most women do not see the situation as they do. Esther Peterson, a former Assistant Secretary of Labor has said:

> "I think it is tragic that there are many women who don't see it. I blame our mass media, magazines, television, and the nature of our community for their failure to stimulate and really to awaken women to those needs."[8]

Radical feminists at the beginning of this last round of feminism, and now, generally recognize that most women do not agree that females are the "second sex" or "second-class citizens," or are discriminated against unjustly. Most women are not radicals; they just want their fair share. National surveys indicated at the beginning of the latest phase of "women's lib" that the majority of men and women thought women had a just position in our society, and are not unjustly discriminated against.[9]

Caricatured Wife

Yet the most radical feminists thought and still think that most women are so oppressed that they do not even recognize their poor situation. Hill describes the case of the *deferential wife*:

> This is a woman who is utterly devoted to serving her husband. She buys the clothes *he* prefers, invites the guests *he* wants to entertain, and makes love whenever *he* is in the mood. She willingly moves to a new city in order for him to have a more attractive job, counting her own friendships and geographical preferences insignificant by comparison. She loves her husband, but her conduct is not simply an expression of love. She is happy, but she does not simply defer to her husband in certain spheres as a trade-off for his deference in other spheres. On the contrary, she tends not to form her own interests, values and ideals; and, when she does, she counts them as less important than her husband's. She readily responds to appeals from women's liberation that she agrees that women are mentally and physically equal, if not superior, to men. She just believes that the proper role for a woman is to serve her family. As a matter of fact, much of

her happiness derives from her belief that she fulfills this role very well. No one is trampling on her rights, she says; for she is quite glad, and proud, to serve her husband as she does. [10]

Hill describes this caricature case as reflecting "the attitude which I call servility." The deferential wife, in other words, is a slave and doesn't seem to know it, according to Hill. And it takes the "enlightened" radical feminists, who are the vanguard of the NEW women, to liberate the poor, ignorant, and oppressed majority of females.

Feminism, A Middle-Class Movement

When analyzing radical feminism, and whether the radical feminists speak for women in general, one should also remember that radical feminists are generally from the middle-class. We define the middle-class herein as families with better than average income, homes, education, cars, etc. They also have elite attitudes, they desire status jobs, they show reverence to higher education, and so forth. According to Aileen S. Kraditor, editor of *Up From the Pedestal*, "Throughout its history American feminism has been overwhelmingly a white, middle-class movement."(p. 15)

Irving Howe in his critique on Kate Millett, the author of *Sexual Politics*, reveals her middle-class biases:

> Most of the time, when she speaks of women she really has in mind middle-class American women during the last thirty years. About the experience of working-class women she knows next to nothing, as in this comic-pathetic remark: "The invention of laborsaving devices has had no appreciable effect on the duration, even if it has affected the quality, of their drudgery." Only a Columbia Ph. D. who has never had to learn the difference between scrubbing the family laundry on a washboard and putting it into an electric washing machine can write such nonsense. As with most New Left ideologues, male or female, Miss Millett suffers from middle-class parochialism.
>
> And more: she suffers from a social outlook which, despite its "revolutionary" claims, is finally bourgeois in character. She writes that "nearly all that can be described as distinctly human rather than animal activity (in their own way animals

also give birth and care for their young) is largely reserved for the male." And again: "Even the modern nuclear family, with its unchanged and traditional division of roles, necessitates male supremacy by preserving specifically human endeavor for the male alone, while confining the female to menial labor and compulsory child care."

These sentences indicate that Miss Millett is at heart an old-fashioned bourgeois feminist who supposes the height of satisfaction is to work in an office or factory and not be burdened with those brutes called men and those slops called children. For one must ask: Why is the male's enforced labor at some mindless task in a factory "distinctly human," while the woman bringing up her child is reduced to an "animal" level? Isn't the husband a "chattel" too? Hasn't Miss Millett ever been told by her New Left friends about the alienation of labor in an exploitative society? And is the poor bastard writing soap jingles in an ad agency performing a "human" task morally or psychologically superior to what his wife does at home, where she can at least reach toward an uncontaminated relationship with her own child? ... In some remarks Miss Millett betrays a profound distortion of values, a deep if unconscious acquiescence not only to the corruptions of the bourgeois society against which she rails but to all those "masculine values" she supposes herself to be against.[11]

Most women have positive attitudes towards being a mother (especially after their first child), but not the middle-class radical women liberators. They devaluate childrearing, and almost everything associated with the housewife, while they give great value to work outside and away from the home.

Poor Kate. Since Millett's earlier radical days she has ended up childless, divorced, alone, embittered, and has been two times hospitalized for mental illness.[48]

How Do Males Oppress Females?

Let's examine the so-called oppression. Generally, the radical feminists don't go into much detail about the oppression. They have

---

[48] **Kate Millett,** *The Loony-Bin Trip,* 1990

no coherent theory of the oppression. They have no genesis of the oppression. And they have no real plan on how the oppression will end because they admit they do not know what is the real woman.

According to the radicals, women have been oppressed for so long that they don't even know what is a liberated woman:

> "The free woman is just being born," says Simone de Beauvoir, in her book, *The Second Sex* (p. 673).
>
> Because of the total oppression, according to Vivian Gornick and Barbara Moran, in their "Introduction" of *Women in Sexist Society* (p. xxv), "there is a whole universe of meaning to be rescued and redefined."
>
> Women have been denied the opportunity and education to discover all their potentialities, and that whatever women's proper sphere turns out to be, women themselves must find it, according to feminists [12]

Lucy Stone, a well-known feminist, pointed out in an 1855 speech the need for women themselves to define the NEW woman:

> "Leave women, then, to find their sphere. And do not tell us before we are born even, that our province is to cook dinners, darn stockings, and sew on buttons." [13]

According to the radical feminists, the oppression is aided and encouraged through various means. The mass media: television, commercial advertising, radio, and books are some principle means of continuing women's oppressed state because these forms of communication demean women by portraying them in limited and even degrading roles.[14] Ironically, 50+ years later, it is the monopolized mass media, controlled by a few elites, that is demeaning mothers by emphasizing the status of career-women over stay-at-home moms. Before it was the demeaning of mothers by stereotyping them as "only" mothers doing monotonous simple-minded chores in the home, while the husband was out in the world doing real work and using his brain. Of course, both of these types of stereotyping were wrong, for being a good mother is not easy and takes a lot of knowledge and ability to do it well, and most work by men outside the home was and still is monotonous. Work is just that, work. Only the few get to do creative work most of their work-life.

Language. The so-called sexist language is also to blame for the continuing oppression, according to the feminist. In city government the use of such words as council*men* seem to imply to the radical feminists that women are excluded as possible city council members. This is a form of subtle oppression according to women liberators. Thus, the radical feminists insist on the use of council*person*. Essays have been written on the so-called sexism of the English language. From *Women in Sexist Society* we read: "As Ethel Strainchamps shows in 'Our Sexist Language,' even our language, praised by philologists for its 'masculinity,' complicates any efforts at redefinition of woman's place by its implicit bias."[15] Of course, these words were not exclusively referring to males, just like mankind is not exclusive to men, for mankind means all humankind. But these words were pointed out by the feminists to make their point, without qualifying that all who heard "councilman," knew it could also include councilwomen; it was merely a shortcut expression.

Institutions. The world's institutions were also felt to be men's means of oppression. The world's institutions kept women down. According to radical feminists, they must be changed and changed radically in order for women to be freed. Betty Friedan put it the following way:

> The changes necessary to bring about that equality were, and still are, very revolutionary indeed. They involve a sex-role revolution for men and women which will restructure all our institutions: childrearing, education, marriage, the family, medicine, work, politics, the economy, religion, psychological theory, human sexuality, morality and the very evolution of the race.[16]

Change Everything. Now that's radical change! Everything must be changed in order for the radical feminists to be freed. And we say "in order for the radical feminists to be freed" because they are the ones who feel enslaved by alleged male oppression. To the radical feminists, it seems better to change *everything*, instead of the radicals moderating their views.

Marriage: The Enemy

According to women liberators, marriage is the biggest means of oppression. Simone de Beauvoir thinks that "the oppression of woman has its cause in the will to perpetuate the family." Freedom comes to women "to the degree in which she escapes from the

2. Grievances and Claims

family."[17] Ironically Ms. de Beauvoir apparently stayed in an oppressive 50-year relationship herself.[49]

Marlene Dixon, in her paper, "The rise of Women's Liberation," says:

> The institution of marriage is the chief vehicle for the perpetuation of the oppression of women; it is through the role of wife that the subjugation of women is maintained. In a very real way the role of wife has been the genesis of women's rebellion throughout history.[18]

According to Aileen S. Kraditor, in her introduction to her book, *Up From the Pedestal*:

> ...The institution of the family itself, as popularly conceived, stands revealed as the obstacle to full sex equality. As long as the man engages in the work of the world and the woman spends a large proportion of her time and energies in the isolated family circle, men will continue to lead government, and professions, and all the other fields that provide us with our criteria of human achievement....
>
> In short, inequality of the sexes still exists because the family structure has remained basically unchanged.[19]

Thus, according to the radical feminists, males oppress females through all the institutions of society, but especially through marriage, for it is the main institution of oppression. It amazes me that such a warped view of reality can even reach the printed page since it is obvious that often it is the woman that wants to get married more so than the man.

Freedom Through Work — or — Change Everything

As we have seen, according to the radical feminists, marriage is the main cause of the oppression of women because the housework, childrearing, and other wifely duties allow little time for the women to achieve. They think that achievement is advancement in traditional male activities. Wifely duties keep women tied to the home and economically dependent on men. And to be economically

---

[49] Jean-Paul Sartre was her "husband," except she had to endure his many open relationships with his lovers and his obvious patriarchal domination. See *A Dangerous Liaison: A Revalatory New Biography of Simone DeBeauvoir and Jean-Paul Sartre* by Carole Seymour-Jones (2009). Although she never legally married him, nevertheless she was oppressed by him.

dependent is terrible, for economic independence is how women will obtain freedom from oppression and gain their autonomy.

According to Betty Friedan:

> For women to have full human identity and freedom, they must have economic independence....Equality and human dignity are not possible for women if they are not able to earn money.[20]

In the words of Simone de Beauvoir:

> Now protected in large part from the slavery of reproduction [by birth control], she is in a position to assume the economic role that is offered her and will assure her of complete independence.[21]

Simone de Beauvoir in her book, *The Second Sex*, says it is through the economic role that women will gain "complete independence" from their slavery under men. Yet Simone says further on in the same book, "we must not believe, certainly, that a change in women's economic condition alone is enough to transform her...." ([21] p.683) She goes on to say that other factors like the moral, social, and cultural ones must also be changed before the NEW woman appears. Furthermore, she indicates that these social changes must come through social evolution, but "no single educator could fashion a *female human being* today who would be the exact homologue of the *male human being*," for she understands that "the forest must be planted all at once" because past socialization cannot be completely erased. She further notes that in an achieved evolutionary state of equality between the sexes, "this equality would find new expression in each individual." This equality Simone later asserts would *not* be "equality in difference" but "differences in equality"([21] p. 688). Simone understands there will always be differences between the sexes (eroticism) because there are innate differences. She notes some of these differences carefully, and only as many as are obvious, for she is probably fearful of being pegged with the so-called antifeminists who call for "equality in difference," or different spheres with equal value — the separate but equal doctrine.

Separate-But-Equal Doctrine

Radical feminists do not believe in the separate-but-equal doctrine. This doctrine is one which claims that women at home are equal to women working at jobs away from home. But radical feminists assert directly and indirectly that work in the home is of little value when compared with work out of the home. The radical feminists do not buy the argument that men are just as dependent on their wife at home as she is dependent on him.[22] Yet here and there they say this in so many words. In the words of Simone de Beauvoir: "The dialectic of master and slave here finds its most concrete application: in oppressing, one becomes oppressed. Men are enchained by reason of their very sovereignty."[23]

Simone is saying that men are oppressed too. But she goes on and relates that men are the *cause* of the oppression of both, thus, the true oppressor. Simone doesn't say specifically that men are dependent on women, for to do so would be to play into the hands of those who call for separate-but-equal spheres. But any half-way analysis of the interaction between the sexes would show the *interdependence* of the sexes. Each is dependent on each other in similar and dissimilar ways.

Working for Your Family v. Working for Others

The separate-but-equal doctrine is rejected by radical feminists merely because they believe that work done in the home is of little value compared to work done outside the home.[24] Women liberators give little value to women's traditional work, ironically, just like the male chauvinists they so despise. The work most females in the world have done and still do is housework and childrearing, which may include caring for the family's children, preparing food for the family, decorating the house, cleaning the house, bookkeeping for and even managing the family estate or business, selling goods at markets, garden work, etc. Housework may also include farm work if they live on a farm. This sounds like a lot of work. But things have changed.

As we mentioned earlier in this book, in 1870, according to the US Census, 15% of women worked outside the homestead. However in 2014 women make up about 45-49%[50] of the workforce in the USA.[51] In 2013, 35% of mothers with children at home in the United States did not work outside the home.[52] The change now is that more women are working for *others* outside their immediate family and are leaving their children with others; they bear the children, but

---

[50] One government table indicates 49%, while another indicated about 45%

[51] Bureau of Labor Statistics — http://www.bls.gov/news.release/empsit.t21.htm ; http://www.bls.gov/news.release/empsit.t09.htm

[52] Bureau of Labor Statistics — http://www.bls.gov/news.release/famee.nr0.htm

they do less of the rearing. As we will see in our motherhood chapter, leaving your children to others while you work is not the best arrangement for the well-being of your children.

Work and Money. Marlene Dixon thinks money is the reason women's work is felt to be of less value:

> Household labor, including child care, constitutes a huge amount of socially necessary labor. Nevertheless, in a society based on commodity production, it is not usually considered even as "real work" since it is outside of trade and the marketplace. In a society in which money determines value, women are a group who work outside the money economy. Their work is not worth money, is therefore valueless, is therefore not even real work. And women themselves, who do this valueless work, can hardly be expected to be worth as much as men, who work for money.[26]

People should know that receiving money for work isn't the standard of true value. Women who don't earn money still make money indirectly through caring for their children and housework. This latter point is one of the reasons wives in many areas own half of the husband's assets. But in any case, even if some women don't own half the family's assets by law, they in fact have full *use of all the family assets*: they sleep in the same bed as the husband, they live in the same house as the husband, they drive the same cars as the husband, they eat the same food as the husband, they take the same vacations as the husband, etc. So in practicality, women own all that their husbands own. And, of course, if women didn't give birth and care for children, we would be a dying race. Each woman not only gives life to her children, but through women we have humanity. This is of *great* value, but radical feminists and many men greatly underestimate it. The true monetary value of childbearing and childrearing can't be valued in monetary terms. Children are the basis of a strong nation: children are the future of any nation; well-reared children equal a strong nation.

To the radicals, women-at-home are "domestic slaves," the "second sex," the "lost sex," the "other," and so on. Therefore radical feminists demand "meaningful work" outside the home. This they call their vehicle to independence.

Meaningful Jobs

The vanguard of the radical feminists not only wanted jobs outside the home for women — they wanted *meaningful* jobs. Shirley Zimmerman, a one time coordinator for women employees of Santa Clara County in California, put it like this:

> Feminists like to have affirmative action jobs, or be in management or hold good clerical jobs. They're not thinking about being bus drivers or mechanics or assessors or analysts. [27]

The radical feminists demand for *meaningful* jobs reflects their middle-classism. They don't want factory jobs; they want jobs with status. They want professional work. Jeanne Clare Ridley, an associate professor of sociomedical sciences, wrote for a government commission on population growth:

> When one examines carefully the rhetoric of the feminists concerning the provision of meaningful work opportunities for women, one is quickly led to the conclusion that it is the professional occupations that are being referred to. This is not surprising since, as already noted, most members of the women's liberation movement are drawn from the well-educated professional classes.[28]

The National Organization for Women (NOW) in their "statement of purpose" of 1966 stated that 75% of the women who work outside the home, work "in routine clerical, sales or factory jobs."[29] But according to the *Statistical Abstract of the United States* [30] for 1974 (p. 350), 72% of the working men were likewise in routine clerical, sales or factory jobs. But this same table shows that 62% of the working women held white-collar jobs while only 41% of the men held white-collar jobs. Even at the beginning of the modern radical feminism, working women still held a greater percentage of the white-collar jobs than men. Of course, they were looking for the well-paying white-collar jobs.

Wage Differences

NOW's statement also indicated that women working full-time make on the average only 60% of what men earn. In 1990, the figure was about 70%, today about 77%, depending on the method of calculation used. In 1988, women with union jobs made 80% of what men in unions made, while non-union women made 72% of what non-union men made. Today, the figures are about the same. This seems like unfair wage discrimination, but this is an oversimplification, for men generally work in more skilled[53] jobs, work more overtime, less men work part-time than women and have more job continuity since they don't stop their careers to have babies. When these things are factored in, women today earn about 91 cents for every dollar men earn according to an Institue for Women's Policy Research's study.[54] How many women do you know work all the overtime they can get? Suzanne Keeler in her paper, "The Future Status of Women in America," wrote:

> Some of the statistics do, indeed, tell a depressing story of lower medium earnings for women in 1971 than in 1958, and of lower shares in professional and technical jobs than in 1939, but these must be analyzed with care. Some of the discrepancies are inflated by facts such as that men moonlight much more than women do, men do more overtime work and less part-time work, and men have fewer job disruptions and greater job continuity.[31]

These facts are still relevant today except that more women are moonlighting. Another reason for the discrepancies in wages is that a greater percentage of the working men, than the working women, belong to unions.[32] On the average, union workers make more than non-union workers in similar jobs. Another factor is that men spend more of their lifetime in the labor force, because of this they have more seniority. But, I personally believe women are still discriminated against in wages. How much I don't know.

Demand For Status Jobs

The middle-classism of NOW is clearly shown in their desire for a bigger piece of the meaningful jobs:

> In all the professions considered of importance to society, and in the executive ranks of industry and government, women, are losing ground. Where they are present it is only a token handful. Women comprise less than 1% of federal

---

[53] counterpoint: is a seamstress less skilled than a carpenter? — the classification of what is a "skilled" job may be another male value determination

[54] Wash. Post, June 5, 2012 — http://www.washingtonpost.com/blogs/wonkblog/post/women-earn-91-cents-for-every-dollar-men-earn--if-you-control-for-life-choices/2012/06/04/gJQAqrHkEV_blog.html

2. Grievances and Claims

judges; less than 4% of all lawyers; 7% of doctors. Yet women represent 51% of the U.S. population. [35]

Today, the percentage of female lawyers and doctors has *dramatically* increased, but as they have increased, lawyers and doctors are loosing some of their former status.[55] Today about 29% of all lawyers[56] and 22-26% of all Federal and State Judges[57] are women and about 28-30% of all doctors are women.[58] This will change even more, for women studying to be lawyers and doctors are approximately 50% of today's classes.

NOW makes a value judgement and says these are the professions "considered of importance to society." NOW then goes on and gives quotes on some of the professions of which women compose a low percentage. The reason, of course, for these low percentages in the 1960's was that women were not studying to be lawyers, or taking management courses in college, or going into medical school.[36] Thirty years later almost as many women are taking these subjects as men. But radical feminists still contest this and tell us there should be more in higher positions, and the reason there are not more is that women are discriminated against. This may have been so to some extent, but today it would be hard to prove. Radical feminists will say that society as a whole teaches women that these professions are for men, thus women have ruled out these fields long before they reach the stage when they can study them. The fact that at one time over 70% of Russian doctors were women might back up the feminist argument to some extent. Yet in Russia, Sweden, China, and other countries that have laws calling for total equality between the sexes, they have a low percentage of women in top government and management jobs.[37]

---

[55] *The Changing Face of Medicine: Women Doctors and the Evolution of Health Care in America*, by Ann K. Boulis and Jerry A. Jacobs, Cornell University Press, 2008

[56] Paula A. Patton, *Women Lawyers, Their Status, Influence, and Retention in the Legal Profession*, 11 Wm. & Mary J. Women & L. 173 (2005), http://scholarship.law.wm.edu/wmjowl/vol11/iss2/3, p. 193

[57] "Women in Federal and State-level Judgeships," *A Report of the Center for Women in Government & Civil Society*, Rockefeller College of Public Affairs & Policy, Spring 2010

[58] *Statistical Abstract* of the USA, 2010, Table 158 (for year 2007)

One-Sided Equality of Women's Lib

However, a point that needs to be made is that radical feminists do not complain that in 2013, 89% of the nursing profession are composed of women or 81% of all elementary teachers are women, or 98% of preschool and kindergarten teachers are women.[59] If they are for equality, then why don't they mention this one-sided discrimination? Again, why don't radical feminists complain that there are too many female librarians (84%) or too many childcare workers (95%), and so forth.[60] Even though men make up 49% of the population and by tradition and some state laws *must* support their families, many occupations are overrun with women. Why aren't radical feminists concerned with rectifying this situation? The short answer: most of these jobs are not well-paying jobs. They want status jobs and well-paying jobs. But as the book by Boulis and Jacobs and the report by Patton referenced (footnotes) on the previous page discuss: when status jobs become women's jobs they loose status. There is a reason for this which we will discuss later in this book and which is also written about with some depth in *The Confidence Code:The Science and Art of Self-Assurance— What Women Should Know*.[61]

Back in 1971 men earned most of the doctorates, approximately 27,500 versus 4,500 for women.[38] Thus, men deserved at that time, generally speaking, 84% of all top job openings in 1971 in all fields where doctorates count. But if radical feminists had their way in 1971 at *least* 50% of these job openings would have gone to women. One suggested "immediate equality" which would have to mean that all openings would be filled with women until the 50% level is reached.[39] But she later called for a "reasonable progress" program somewhere between "gradual improvement" and "immediate equality" because she felt the status quo would resist too much if the "immediate equality" program was initiated.

---

[59] http://www.bls.gov/cps/cpsaat11.pdf

[60] http://www.bls.gov/cps/cpsaat11.pdf

[61] Katty Kay & Claire Shipman, *The Confidence Code*, HarperBusiness, 2014: men are "wired" for confidence; women are not, and what can be done to help alleviate this. Confidence and status are related.

2. Grievances and Claims

Affirmative Action

1. When feminists speak of affirmative action, they ignore any professional occupations where women are in the majority. Are they for equality in these areas?

2. Many radical feminists insist that all professions with low percentages of women be brought up to the 50% level even though few women enter these fields because they lack interest, or because they are more concerned with motherhood, or because it takes too long for them to enter these fields due to the amount of preparation, or there may be some physical reasons for them not entering (strength), or because there may be some mental reason for them not entering (lack of spatial ability), or because women are more people oriented, or there may be other reasons, such as confidence, for women not entering some professions.[62] But through such programs as affirmative action, the few women who are seriously interested in male occupations like engineering have an unfair advantage over the males in these fields, since these few women can get jobs easier because of the pressure on employers to hire women. Thus these few women have *more freedom* than men.

3. Women have babies and universally nourish and care for infants most of the time, whereas men do not have babies and universally do not nourish and care for infants most of the time. As we point out in our Chapter 3 on sex roles, radical feminists can only point to a few rare instances where men take care of infants such as the one supposed case of the Mountain Arapesh that the pop-anthropologist and bisexual Margaret Mead reported on. (see Chapter 3) But in this case, the women cared for the young children the *greatest* proportion of the time. The mountain Arapesh women carried their infants whenever they traveled, and spend most of the day with their children. A father cared for the children mostly when the mother was in menstrual seclusion (if he didn't have a second wife), or when the mother had to perform some womanly duty like getting some greens for the evening meal.[40] If one reads Margaret Mead's books carefully, one sees they are not all that radical feminists

---

[62] see Chapter 3 & 4 of this book and see the many discussions about confidence in *The Confidence Code,* by Katty Kay & Claire Shipman, Harper Business, 2014

claim. Also, all day-care centers — including those in the Israel's Kibbutzim, China, Russia, and, of course, the USA, are mostly staffed by women.[41]

4. Thus, since women have the profession of motherhood, and men do not, it would be a form of sex discrimination to make all the other professions for men take in up to 50% women.

Affirmative Action Program. The feminist Affirmative Action program thus unfairly gives some women advantages in competition with men through unjust quota systems, especially in the few status jobs. The Affirmative Action plan in the USA was set up to increase the utilization of women in various professions and job categories. [42] But curiously, as we have noted, in such professions as nursing and child care, the affirmative action programs are not being used to balance these professions. Only in the well-paying male dominated occupations are radical feminists trying to balance the numbers of men and women.

As we will see in later chapters of this book, women have certain traits that make them better suited for certain jobs, such as editors, hand finery jobs, childcare, color coordination,[63] etc. In these, they should be well-represented,[64] but with fairer wages.

Sex roles are the way they are, mostly because of the biological differences between men and women (see Chapter 4). However, now women are expected to work full-time *and* be mothers, which is truly unfair. To radical feminists, cultural conditioning is the main cause for sex/gender roles. Thus, we enter the great bio-cultural debate.

---

[63] A portion of women have an extra color cone and can perceive millions of more colors than other women and men; they have tetrachromat vision.

[64] but not by some artificial affirmative action program, but by hiring the best candidate, of which in fields like editing, women would make better candidates on average since they are better at details than men.

2. GRIEVANCES AND CLAIMS

References for Chapter 2

[1: ch 2] is 95 of the Bibliography list
[2: ch 2] is 155
[3: ch 2] Numbers 27:3-4, *New English Bible*
[4: ch 2] pp. 86, 88-89 in 42
[5: ch 2] 3rd book, chapter 12 in 5
[6: ch 2] p. 97 in 196
[7: ch 2] p. 109 in 126
[8: ch 2] pp. 218-219 in 52
[9: ch 2] pp. 18-19 in 80
pp. 61-63 in 199
[10: ch 2] p. 89 in 83
[11: ch 2] p. 183 in 84
[12: ch 2] p. 17 in 95
[13: ch 2] p. 72 in 95
[14: ch 2] p. 219 in 52
[15: ch 2] p. xxv in 67
[16: ch 2] p. 30 in 57
[17: ch 2] p. 82 in 42
[18: ch 2] p. 189 in 46
[19: ch 2] pp. 22-23, 24 in 95
[20: ch 2] p. 30 in 57
[21: ch 2] p. 112 in 42
[22: ch 2] pp. 11-12 in 95
[23: ch 2] p. 454 in 42
[24: ch 2] p. 23 in 95
[25: ch 2] see latest Almanac
[26: ch 2] p. 189 in 46
[27: ch 2] p. 28 in Working for Women. In "California Today," *San Jose Mercury-News*, Sunday May 5, 1974, pp. 14ff
[28: ch 2] p. 300 in 144
[29: ch 2] found in 130
[30: ch 2] p. 350 in 194
[31: ch 2] p. 276 in 90
[32: ch 2] p. 241 in 193

[33: ch 2] p. 398 in **194**
[34: ch 2] p. 398 in **194**; in **197**
[35: ch 2] found in **130**
[36: ch 2] p. 133 in **193**
[37: ch 2] in **54**; **178**; **183**
[38: ch 2] p. 139 in **194**
[39: ch 2] found in **81**
[40: ch 2] p. 1439 in **111**; pp. 54-55, 114-115 in **109**
[41: ch 2] found in **98**; **183**
[42: ch 2] pp. 91-92 in **148**

3. Sex Roles
Biology v. Culture

Sex Role Development

As far back as 1935 Margaret Mead wrote about the need for freer sex roles.[1] Mead believed that women should be able to take part in more activities usually associated with men. Others before her called for more freedom from sex role stereotypes.[2] Sex role stereotypes have to do with our expectations about the behavior of males or females: generally, men work outside the home, do the heavy physical work, fight the wars, dominate most if not all spatial ability jobs (mathematicians, engineers, architects) and control the most powerful institutions; women bear and nurse the babies, care for the young, work in or near the home, work in family businesses, do the family cooking, work outside the home to help support their family in hard times, etc. When they work outside the home for others they work in more occupations that deal with children, interrelationships between people and jobs that need good verbal abilities.

This has changed somewhat in industrialized countries since Margaret Mead's time through the help of automation and computerization of many jobs and the pressure of economics. Today about 51% of all women in the world participate in the labor force,[65] but at the same time they are still doing work at home while caring for their children and cooking, etc. Is this better?

Thus, in the last several decades many traditional sex roles have been "successfully" challenged [3]: Not only are more women working outside the home, but more women are working who have children, **because:**

1. of the equality dogma taught in the schools and media,
2. men thought monetarily it would be nice to have a working wife (neglecting to see the negative aspects)
3. of the desire for corporations to keep the cost of labor low ("we can pay them less")
4. domestic and industrial devices (dishwashers, microwaves, frozen food, etc.) and job-related devices (computers, forklifts, etc.) made it possible, etc.

---

[65] "Women in Labour markets," International Labour Office, Geneva, 2010

The movement has thus "liberated" women. So "successful" was the movement that in just a little over four decades after the 1960s, most women are working outside the home as well as in the home. In 2005, about 71% of women in the USA with children under 18 were working outside the home. But is this a success story? Women and children have paid an unfair price for these changes.

Let's now examine the evidence for and against environmental or social influences pertaining to sex/gender role development. In Chapter 4, we will look at the real biological differences that help produce the different sex/gender roles.

Margaret Mead and Cultural Conditioning

According to the well-known anthropologist Margaret Mead:

> All known human societies recognize the anatomic and functional differences between males and females in intricate and complex ways; through insistence on small nuances of behavior in posture, stance, gait, through language, ornamentation and dress, division of labor, legal social status, religious role, etc.[8]

But Mead goes on to indicate that in other cultures besides ours the sex roles are more *flexible*. She quotes evidence from primitive cultures — that is cultures that are not large in numbers or great in ideas or great in wealth. Mead believed, like other radicals, that sex roles and behavior are caused primarily by cultural conditioning or socialization.

In *Sex and Temperament*, Margaret Mead described the supposed proof that sex roles are culturally derived. Others such as Nancy Chodorow, the author of *Being and Doing: A Cross-Cultural Examination of the Socialization of Males and Females*, mention and relate Mead's studies like this:

> Cross-cultural research suggests that there are no absolute personality differences between men and women, that many of the characteristics we normally classify as masculine or feminine tend to differentiate *both* the males and females in one culture from those in another, and in still other cultures to be the reverse of our expectations.

> Margaret Mead's studies describe societies in which both men and women are gentle and nonaggressive (the Arapesh);

in which women dislike childbearing and children and both sexes are angry and aggressive (the Mundugumor); in which women are unadorned, brisk and efficient, whether in childrearing, fishing, or marketing, while men are decorated and vain, interested in art, theater, and petty gossip (the Tchambuli)....[9]

Nancy Chodorow is speaking about personality, or temperament differences in *certain few selective* societies, for she realizes that in most cultures male and female behavior does conform to our traditional expectations:

This is not to claim that within most cultures, male and female differences do not generally conform to our traditional expectations. George Murdock's and Roy D'Andrade's data on the division of labor by sex indicate that most work is divided regularly between men and women, along conventional lines. Men's work, for instance, is strenuous, cooperative, and...may require long periods of travel; women's work is mainly associated with food gathering and preparation, crafts, clothing manufacture, child care, and so forth.[10]

Therefore what Nancy Chodorow is saying, in her own words, is that: "Cross-Culture research *suggests* that there are no *absolute* personality differences...."[11]

Some like Chodorow say that the studies like Margaret Mead's *suggest* that many of the differences in sex behavior are culturally determined, while at other times they claim Mead's work *proves* that gender behavior is mainly determined by cultural conditioning. They often forget that Mead's work, even if it were true, can only *suggest* that gender behavior differences are determined by cultural conditioning. In Chodorow's own words: "This essay refutes the claim for universal and necessary differentiation, and provides an explanation based on a comparison of cultures and socialization practices to account for such differences where and when they occur." [12] She forgets reality and projects her bias here.

Notice how Mead puts it in her book, *Sex and Temperament*. After describing three cultures (Arapesh, Mundergumor, and Tchambuli) where sex behavior is claimed to be of a different order than traditionally expected, Mead concludes:

These three situations *suggest*, then, a very definite conclusion. If those temperamental attitudes which we have traditionally regarded as feminine — such as passivity, responsiveness, and a willingness to cherish children — can so easily be set up as the masculine pattern in one tribe, and in another be outlawed for the majority of women as well as for the majority of men, we no longer have any basis for regarding such aspects of behavior as sex-linked. And this conclusion becomes even stronger when we consider the actual reversal in Tchambuli of the position of dominance of the two sexes, in spite of the existence of formal patrilineal institutions. The material *suggest* that we may say that many, if not all, of the personality traits which we have called masculine or feminine are as lightly linked to sex as are the clothing, the manners, and the form of head-dress that a society at a given period assigns to either sex.... *We are forced to conclude that human nature is almost unbelievably malleable*, responding accurately and contrastingly to contrasting cultural conditions. The differences between individuals who are members of different cultures, like the differences between individuals within a culture, are *almost entirely* to be laid to differences in conditioning, especially during early childhood, and the form of this conditioning is culturally determined. Standardized personality differences between the sexes are of this order....[13] (my emphasis)

Margaret Mead first says that three different cultures *suggest* the malleability of sex roles, then she concludes that cultural conditioning determines sex behavior differences "almost entirely," "we are forced to conclude that human nature is almost unbelievably malleable." She believes sex roles are *forced* on the sexes:

> In the division of labour, in dress, in manners, in social and religious functioning — sometimes in only a few of these respects, sometimes in all — men and women are socially differentiated, and each sex, and a sex, forced to conform to the role assigned to it.[14]

Moreover in Mead's thinking, as with many radical feminists, if society recognized individual differences instead of conditioned sex differences:

> It might abandon its various attempts to make boys fight and to make girls remain passive, or to make all children fight, and instead shape our educational institutions to develop to

the full the boy who shows a capacity for maternal behavior, the girl who shows an opposite capacity that is stimulated by fighting against obstacles. No skill, no special aptitude, no vividness of imagination or precision of thinking would go because the child who possessed it was of one sex rather than the other.[15]

Exceptions Rule. Mead as well as others base their thinking on certain cultural studies which seem to them to manifest the extreme malleability of sex behavior. Of course these certain cultural studies are *exceptions* to the universal gender behavior differences, and besides that they were biased and failed studies as we will see, but this doesn't stop radical feminists.

For them the exceptions rule. To them, we must change all societies to fit their idea of equality. And their way to do this is through social conditioning. Thus, radical feminists call any thought "sexist" if it manifests any idea of difference between the sexes. If there are differences, they are culturally derived. Women liberators only concede the very obvious — the genitals and breasts, and downgrade the other differences.

In 1971, before the Committee on the Judiciary in the House of Representatives, Representative Don Edwards, concerning the Equal Rights Amendment, asked:

> I refer you back to Margaret Mead who more than 50 years ago, who in her studies of the New Guinea nation and tribes, came to the conclusion that sex roles result from social learning rather than biologically inherited tendencies.
>
> In other words she is saying that it is cultural, not biological. I am sure you will recall that among the Mundugumor Tribe of New Guinea both the sexes acted about the same; they were hostile, aggressive and violent, qualities that we generally associated with men in this country.
>
> In the Tchambuli Tribe, the women were practical, domineering and aggressive; the men sensitive, artistic, emotional, and dependent.
>
> In other words, the roles were theoretically reversed.
>
> In the Arapesh, the men had the general temperament which some people think is feminine.
>
> Do you believe that men and women, except for the one or two obvious differences, the differences of sex and the fact that

women do have babies, do you think that men and women are really about the same?

Witness Number Two Answers Edwards: I think it is the consensus that men are sperm donators, women are baby incubators, and all the rest of it is the result of the socialization process.66

Mead's Superficial Studies

Arapesh, Mundugumor, and Tchambuli. Let's take a deeper look at Mead and her writings on the Arapesh, the Mundugumor, and the Tchambuli people. Mead published her work, *Sex and Temperament*, in 1935. In 1936 Lewis M. Terman and Catharine Cos Miles published *Sex and Personality: Studies in Masculinity and Femininity*.[16] In this work Terman and Miles had some criticism of Mead which we will pass on here:

> In a recent treatise Mead has presented a mass of descriptive evidence favoring the extreme environmental hypothesis for the causes of sex difference in personality....
>
> That Mead's contribution offers impressive evidence of the modifiability of human temperament will be readily conceded, but we are by no means convinced that the case for nurture is as strong as a casual perusal of her book would suggest....It is not to be supposed that the field anthropologist, any more than the psychologist, is immune to error in such estimates [of the degree of masculinity or femininity]; indeed, because the groups under observation by him belong to an alien culture, and because his command of the tribal language is almost invariably limited, the anthropologist who attempts to rate the masculinity or femininity of behavior in a primitive tribal group labors under tremendous disadvantages.(p. 461)

Mead, herself, adds to this thought train of Terman and Miles in one of her writings:

---

66 (p. 493, *Equal Rights for Men and Women 1971*, U.S. Government Printing Office, Washing: 1971; Witness number two was from the George Washington University Women's Liberation)

3. SEX ROLES

> Explicitly, as a matter of training we [anthropologists] send our students out to remote and exotic peoples, where they will be exposed to ways of behavior quite different from our own, so different in fact that no effort of the mind will work that simply redefines the new ways in terms of the known old ways. Living among a people all of whose ways are alien, anthropologists make many adjustments. We learn to speak, or at least to hear and think, in quite different languages, in which there may be many genders or none, in which words may be put together in ways that defy all our attempts to fit them into our familiar grammatical categories....[17]

Anthropologists like Mead go to areas of the world with customs and languages that are unfamiliar to them. It is difficult for us to observe and correctly judge our own society, even our own family, neighbors, and friends let alone alien societies with radically different languages.

Mead's Superficial Foreign Language Knowledge

Mead spent only seven months with the Mountain Arapesh, three and one-half months with the Mundugumor people, and only several months with the Tchambuli people.[18] Yet in these few months she supposedly learned their languages and enough of their customs to make judgments on these alien peoples. But did she?

Pidgin English. Margaret Mead claims that primitive cultures are simplistic enough to learn much about them in short periods of time *vis a vis* such complex societies as America where it is much more difficult.[19] But this may be so to Mead merely because she knew the foreign languages only superficially. Mead has said: "In all cases the language was learned, a base was established in a native village, and one village was intensively followed and studied."[20] But how were the languages learned?: "In Manus we had to analyze the language, *using pidgin English as the interpreting language*, and this was true also of Arapesh, Mundugumor, and Tchambuli ...."[21, emphasis mine] Thus, she used an interpreting language, Pidgin English "from a Manus-speaking schoolboy with an understanding, although hardly any speaking knowledge, of English...."[22] In the case of the Tchambuli people she admits, "They speak a difficult Papuan language...."[23] This is preposterous. Only the pop-educated can exalt a study based on pidgin English.

Mead's Preposterous Study. Mead, we are supposed to believe, spends a few months in alien *cultures*, studies one tribe in each culture, and in this period learns their difficult languages (made even more difficult by their many times unknown histories), ascertains their customs, and through her training unbiasedly perceives their true gender behavior. Then because of Mead's dubious study of these primitive cultures with small populations (about 85 people of the Mountain Arapesh, about 500 people of the Tchambuli, and about 1000 people of the Mundugumor)[24]; and because of other such studies by Mead and other anthropologists, we must acknowledge the *possibility* of the malleability of sex roles and therefore agree and accept the mass conditioning of the sexes into "true equality" as defined by the radical feminists. This kind of reasoning is warped. It should be called what it is: a biased, mystical, almost cultic exercise in myth-making. And the detrimental actions by our government, schools, and institutions that push this reasoning on us come from the old-radical feminism.

Other Points Against Mead's Work

Biased Work

(1) Mead's outlook *before* she studied these tribes was biased. Terman and Miles wrote in a footnote about her bias back in 1936, in their book, *Sex and Personality*:

> In the specific case at hand, it is no reflection upon Dr. Mead to call attention to the fact, verifiable by examination of her earlier writings, that she entered upon her study of sex and temperament with definite leanings toward the environment hypotheses in the interpretations of human behavior patterns. If the composite verbal pictures of her three New Guinea tribes had been sketched by a equally competent observer of different bias, there is no way of knowing how they would have differed from the dramatic contrasts presented; we can only be certain that they would have differed. (p.462)

Parenthetically, as to whether sexual behavior was culturally or biologically determined, Terman and Miles were unsure. (pp. 451, 460) Margaret Mead went to school in an age during which such books as *Patterns of Culture*[25], by Ruth Benedict, were published. In the first chapter of this 1934 book we read:

> The life history of the individual is first and foremost an accommodation to the patterns and standards traditionally

handed down in his community. From the moment of his birth the customs into which he is born shape his experience and behaviour. (p. 18)

Mead's and Benedict's Homosexuality. Probably, the main reason for Mead's bias was that she heavily leaned towards homosexuality, thus she was trying to propagate her own biased belief to make her peculiar behavior more acceptable:

> Margaret Mead, who died in 1978, and Ruth Benedict, who died in 1948, were bisexuals. They were lovers, but each had been married—Mead to at least three husbands, Benedict to one. Rumors of the Mead-Benedict affair were hushed around Columbia University circles in the 1930's, and it was not well known in other academic centers. An account of the affair appears in a Margaret Mead biography written by her daughter, Mary Catherine Bateson. It's titled "With a Daughter's Eye: A Memoir of Margaret Mead and Gregory Bateson."[67]

Was Mead rationalizing her homosexuality when she wrote:

> Where 'logic' is regarded as male, and 'intuition' as female, little girls with a capacity for logical thought may be pushed toward [sex role] inversion as a *preference*, for a socially perceived difference between expectations for men and women, or as an identification with a father whose mind corresponds to the cultural stereotype.[25a]

When I began researching sex differences I had no idea Mead was sexually ambivalent; now we can all examine Mead's work better with this knowledge of her ambivalence in her sexuality.

Poorly Documented and Vague

(2) A second point against Mead's work is that careful analysis of Mead's material is most difficult. Notice what the authors, Talcoot Parsons and Robert F. Bales, of *Family, Socialization and Interaction Process* say about this:

---

[67] *Parade*, January 27, 1985, p. 4; *With a Daughter's Eye: A Memoir of Margaret and Gregory Bateson*, by Mary Catherine Bateson, William Morrow & Co, 1984

> She [Mead] often states what does *not* appear among the "X" or "Y" — in comparison usually to the American family — than what does appear. The system reference continually shifts, which is confusing from the present point of view, if not from her own. And if she had realized what she was saying in certain cases, she certainly would have looked twice. Such, for instance, is the case for the Arapesh.[26]

Parsons and Bales go on to show some weaknesses in Mead's assertions. It should be mentioned that these authors state on the same pages that *they*, "must accept the position that the predominant pattern of [sex] differentiation is not constitutionally inherent...."

Other authors like John Nash in his textbook, *Developmental Psychology: A Psychobiological Approach*(p. 206ff), and Ralph Piddington in his book, *An Introduction to Social Anthropology* (Vol. 2, p. 632ff) also write about Mead's inferior study. From Piddington's book we quote:

> In presenting her material, Dr. Mead constantly asks us to accept at their face value her own formulations, in a different language, of the attitude of individual natives and, more questionable still, of the culture as a whole. To describe, for example, the Mundugumor as "a society that counts loyalty to be a stupid disregard of the real facts about the essential enmity which exists between all males" implies a degree of abstract formulation not usually found in primitive society, and the manner in which such concepts are used without any attempt to correlate them with real human behavior suggests that the ethnographer merely hypostatized her own impressions of native life. Such impressions may be wholly or partially correct or may be entirely misleading. The absence of detailed and comprehensive documentation makes it impossible to decide, and the position was well summarized by Nadel: "One cannot help feeling dubious and a little helpless in face of this ambitious theory of sex, based as it is on so meager a material, and so simplifying and elusive an interpretation. The complete lack of exact definitions and tangible criteria in Dr. Mead's book turns this most intricate problem of human psychology into a subject of novelistic exercise rather than of scientific examination."
>
> In spite, however, of the lack of documentation, we should be more inclined to accept Dr. Mead's impressions if there were indications that the facts had been comprehensively

observed and objectively considered. Unfortunately, so far as empirical material is cited at all, it is always from a specific point of view, and it is impossible to escape the conclusion that facts which might lead to an entirely different type of interpretation have been overlooked, while we find the same facts interpreted in contradictory ways in different contexts, in order to fit them into the pattern. (pp. 633-634)

Based on my own study of Mead's work and on other authors' opinions noted above, it would be very careless of anyone to base any theory on her dubious material. Mead's work is nothing but a fictional work. In her twisted "facts" she projects to us her hidden agenda.

Broken Cultures

(3) A third point against Mead's work is that the three societies studied by Mead in *Sex and Temperament* were broken cultures, according to her own report:

The **Tchambuli people**, when studied, had seven to eight years previously just been brought back to their ancestral land by the government because they had fled their land in fear of the warlike Iatmuls. The Tchambuli people were in the process of trying to rebuild their broken culture when Mead was studying them.[27] Such a broken primitive culture, that has been transplanted or restricted from its native land, is filled with problems, as indicated by the Ik tribe's experience.[28] Such problems do affect the broken culture's behavior.

The **Mundugumor people** when studied "presented the picture of a broken culture. Ceremonials were infrequent; a large number of men were away at work [this explains why the women had to fish, climb coconut trees, and their seemly aggressive behavior — their men were away], only a few of the first group of recruits to go away had come home."[29]

The **Mountain Arapesh** when studied had a big problem obtaining food, "Conditions of life are hard, food is scarce, the protein intake is very inadequate, and members of the tribe who live under primitive conditions exist well below their potential energy output...." [30] This helps to explain why the men of this culture seemed more co-operative, nonaggressive, had little sex drive, and so forth. It is well known that men pound per pound must have more protein, and more energy foods because they burn more energy pound per pound.[31] And it is also known that semi-starvation leads to loss of sex interest.[32] Another possible reason for the male Arapesh's lack of passionate desire for their mates,[33] is that their wives are betrothed to them at the age of five at which time she moves into his household as one of the family.[34] As recent research in kibbutz communities indicate, when unrelated boys and girls grow up in the same dwelling place, amorous relationships are rare in later life, even though such behavior was not looked down upon in the kibbutz communities.[35] The intimacy of rearing the child bride in the same household as her future husband, thus, has a diminishing effect on the husband's desire for her. Furthermore, in a cultural setting where males don't receive enough protein, they are less muscular, and thus strength differences disappear. Hence, the males would be less physically aggressive in such a situation.

"Matriarchal" Societies

Other works also indicate that many of the supposed female-dominated societies are *broken societies* or superficially examined societies. From John Nash in *Developmental Psychology*:

> The existence of matriarchal societies has also been taken as evidence for the cultural nature of sex differences, especially in matters of dominance and leadership. The fact is that matriarchal societies are both unusual (and hence atypical) and also involve an apparent rather than a real dominance by women (Piddington, 1950). The Iroquois is often cited as an example of matriarchal society.... Because many men died young and active in battle, there was reduced survival of wise old men to run the affairs of the tribe at home, and of necessity a great deal of authority was delegated to women. However, the important business of this society was making war, and this remained firmly in the hands of the men (Piddington, 1950). Similar instances can be seen today in various parts of Europe, as in Scottish fishing communities. Here also there is what could be described as a matriarchy and for somewhat the same reason. Each year the men spend several months away from home as they follow the shoals of herring around the coast, and the women remain at home in charge of affairs there. They administer the finances, and frequently order the gear and supplies that are needed for the boats. One might interpret this as a feminine control, but again the central activity of the community is held by the

men, who maintain the boats and take them to sea. (p. 206-207)

I want you to note John Nash's statement's pertaining to the males still doing the "important business" of society. This is the bias of males: That whatever males do is *important*. It is highly subjective. We will speak more on this subject later in this work. But for now I just want you to note this bias.

Traditional Roles Still Prevailed

(4) A fourth point against Mead's work is that traditional roles still prevailed in the three cultures described by Mead:

- Women cooked, did the housework, nursed the children, and cared for the children most of the time.[36]
- Looking past Mead's rationalizations, men still had as much authority and dominance as in most other societies, considering the nature of these *broken* societies.[37]
- The men worked outside and away from the homes *vis a vis* the women's work in and around the home.[38]

Traditional Roles are in All Other Primitive Societies.

(5) A fifth point against Mead's work is that in most, if not all, primitive societies including the ones studied by Mead the following is true, according to authorities on the subject:

> The division of labor is along traditional lines: men work away from the homes, they do the heavy work, they do the hunting, the war fighting, the fishing; women work close near the homes, care for the children, cook for the family, and so forth.[39]

> Men ultimately control or dominate the societies through the most powerful institutions.[40]

Persons in Reversed Roles were Maladjusted

(6) A sixth point against Mead's work is that in her study, the most *maladjusted* persons were those of the sex that were, according to Mead, in reversed positions or roles from the world's traditional role forms.[41] This indicates that traditional sex roles are more comfortable to each sex's biological and psychological make-up.

Hormonal Levels not Studied

A final point against Mead's work is that in cross-culture studies various biological sexual factors such as the sex hormonal ratios between androgens and estrogens have NOT been ascertained in most, if not all cases.[42] Thus, the radical feminists cannot prove their claim that culture conditioning, as opposed to biological factors, causes sex differences without proof that the persons studied in these primitive cultures had normal hormonal ratios. Those studied with seemly offbeat sexual behavior may, for all we know, have been freaks as far as sex hormonal ratios go, or as other sex factors go. Mead knew of this problem:

> The second difficultly is that nowhere in primitive studies do we have determination of chromosomal, gonadal, and hormonal sex, or of somatotype constitutions of the individuals in the society whose sex behavior has been studied.[68]

In concluding, contrary to what radical feminists assert, cross-cultural studies by Mead and others do not prove that the main sex roles and behavior of males and females are culturally determined. The supposed exceptions of Mead and others, when analyzed are not clear exceptions but follow the traditional patterns of the world. And these traditional patterns are caused mainly because of the absolute functional differences between the sexes, and only secondly, because of other relative differences between the sexes (see chapter 4).

Experimental Societies' Disasters

Various groups and societies, like radical communes, the communistic Russian society, and Israel's kibbutzim help refute the radical feminists claims that most sexual behavior is culturally conditioned. Show us the new society where "equality," as radical feminists define it, exists. Communes of all kinds have failed and continue to fail. If any have succeeded, you can bet that the radicals would bring it to our attention again and again.

---

[68] Mead, in 111, p. 1437, footnote 3; Mead did note a single exception to this

3. Sex Roles

Russia's Equality

The old Soviet Russia has failed in her mass effort to create "equality" between the sexes. They at first tried to, in effect, abolish the family. From Nicholas S. Timasheff, in his paper, "The Attempt to Abolish the Family in Russia," found in *A Modern Introduction to the Family*, we see that divorce was made easy for a period of time after Russia's revolution:

> The decrees of December 17 and 18, 1917, permitted every consort to declare that he wanted his marriage to be canceled. No reasons were to be given to the board... In addition to this, incest, bigamy, and adultery were dropped from the list of criminal offenses. Abortion was explicitly permitted by the decree of November 20, 1920... If one of the consorts was absent, he or she was notified by a postcard....The anti-family policy was crowned by partial success: around 1930, on the average, family ties were substantially weaker than they had been before the revolution. But this partial success was more than balanced by a number of detrimental effects unforeseen by the promoters of Communist experiment. About 1934, these detrimental effects were found to endanger the very stability of the new society and its capacity to stand the test of war. Let us review these effects....
>
> ... ominous decrease of the birth rate... in 1934, in the medical institutions of the city of Moscow, 57 thousand children were born, but 154 thousand abortions were performed.... In 1934, in 100 marriages there were 37 divorces..
>
> The dissolution of family ties especially of the parent-child relations threatened to produce a wholesale dissolution of community ties, with rapidly increasing juvenile delinquency as the main symptom... crimes in which the sadistic joy of inflicting pain.... Sometimes the schools were besieged by neglected children; other times gangs beat the teachers and attacked women, or regularly fought against one another.
>
> Finally, the magnificent slogans of the liberation of sex and the emancipation of women proved to have worked in favor of the strong and reckless, and against the weak and shy. Millions of girls saw their lives ruined by Don Juans in Communist garb, and millions of children had never known parental homes... The unfavorable development had to be stopped and to achieve this the government had no other

> choice but to re-enforce that pillar of society which is the family....
>
> And actually in 1935, the Soviet government started to prosecute men for rape who 'changed their wives as gloves,' registering a marriage one day and divorce the next....
>
> In the official journal of the Commissariat of Justice these amazing statements may be found:
>
> The State cannot exist without the family. Marriage is a positive value for the Socialist Soviet State only if the partners see in it a lifelong union. So-called free love is a bourgeois invention and has nothing in common with the principles of conduct of a Soviet citizen. Moreover, marriage receives its full value for the State only if there is progeny, and the consorts experience the highest happiness of parenthood.([43] pp. 55ff)

Thus, when the population growth rate of Russia took sharp drops after their revolutionary effort to make the sexes equal, birth-control clinics were limited, sale of contraceptives slowed, abortions became more prohibited, and so forth. When divorces increased, they were made harder to obtain.[43]

Women liberators can't turn to the large number of working women in Russia as something achieved through new equal social conditioning, for Russia lost about 20,000,000 men in the second world war. Russia greatly needed her women workers: "Indeed, it may be fair to say that the Soviet economy and Soviet society, at least until now, could not operate... without the labor provided by women."[44] And because of this Russia used much propaganda to keep and gain women into the work force.[45]

In 1959 four-fifths of the Soviet working women were occupied in production, and less than one-fifth in services like teaching, science, and medical, and they were under represented in management and top government jobs.[46] Furthermore the Soviet men did not help most women workers with the housework, thus these women perform almost double the work of their husbands.[47] Some equality! And when Soviet women achieve a large percentage of a profession — over 70% of the doctors at one time in Russia were women — they didn't even gain status since doctors in the Soviet Union were considered a low status profession.[48] Even Karen DeCrow, a former president of the National Organization for Women (NOW), thought equality had not arrived in the old Soviet Russia.

According to DeCrow, "Equality is not part of their ideology. In their heads, women are different. There is a sexist attitude to women...." [49]

The "Equality" in Israel's Kibbutz

Israel's kibbutz was another example where social conditioning failed. Dan Leon, the author of *The Kibbutz*, tells us that one of the kibbutz's goals was that, "equal rights would be granted to all...this would include equality between men and women."[50] Dan Leon continues:

> The emancipation of the woman and complete equality of the sexes was one of the most important goals of the kibbutz from its inception....
>
> The determination to free the woman from their traditional role as dependent upon the master of the house or breadwinner, and from exclusive subjugation to the household and to the children, was one of the sources of communal education. The communal nursery would open the road to real and not only formal equality. The woman would be free to do equal work and become an equal member of society, sharing equally in its obligations and privileges. This was both an economic need and an integral part of the kibbutz vision.
>
> The vision has come to life in the kibbutz. As a wife, the woman is no longer economically dependent upon her husband, and as a mother no longer tied down remorselessly to her children. She is an equal member of the community, enjoying the complete security it offers her and her family, and the community has removed all those barriers which might prevent her from playing her equal role in every field of its life. Yet the realization of this dream has probably been accompanied by deeper problems and a deeper consciousness of the disparity between the hope and the reality than in any other aspect of kibbutz life. Though, as elsewhere in kibbutz life, light and shadow exist side by side, it would be dishonest to deny that some of the problems of the woman in the kibbutz still await their complete solution. [51]

Problems with Women in the Kibbutz

There is a "problem with women" in the kibbutz.[52] According to M.E. Spiro, the author of *Kibbutz: Venture in Utopia*, the women of the kibbutz have poor morale:

> One source of the woman's morale is that many women are dissatisfied with their economic roles....When the vattikim [original settlers] first settled on the land, there was no sexual division of labor. Women, like men, worked in the fields and drove tractors; men, like women, worked in the kitchen and in the laundry. Men and women, it was assumed, were equal and could perform their jobs equally well. It was soon discovered, however, that men and women were not equal. For obvious biological reasons, women could not undertake many of the physical tasks of which men were capable; tractor driving, harvesting, and other heavy labor proved too difficult for them. Moreover, women were compelled at times to take temporary leave from that physical labor of which they were capable. A pregnant woman, for example, could not work too long, even in the vegetable garden, and a nursing mother had to work near the Infants House in order to be able to feed her child. Hence, as the Kibbutz grew older and the birth rate increased, more and more women were forced to leave the "productive" branches of the economy and enter its "service" branches. But as they left the "productive" branches, it was necessary that their places be filled, and they were filled by men. The result was that the women found themselves in the same jobs from which they were supposed to have been emancipated — *cooking, cleaning, laundering, teaching, caring for children*, etc.

> ...What has been substituted for the traditional routine of housekeeping...is more housekeeping — and a restricted and narrow kind of housekeeping at that. Instead of cooking and sewing and baking and cleaning and laundering and caring for children, the woman in Kiryat Yedidim cooks *or* sews *or* launders *or* takes care of children for eight hours a day....This new housekeeping is more boring and less rewarding than the traditional type. It is small wonder, then, given this combination of low prestige, difficult working conditions, and monotony, that the chavera [female member of the Kibbutz] has found little happiness in her economic activities.[53]

3. Sex Roles

Traditional Family Tendencies in the Kibbutz

The women of the kibbutz often became proponents of "familistic tendencies." According to Menachem Gerson, the writer of the paper, "Women in the Kibbutz":

> ...Age-old problems of women persist in the kibbutz. Many women of the founder generation are dissatisfied and disillusioned. Now middle-aged and older, they find that many of their once-meaningful jobs have become too strenuous. Kitchen and dining room, laundry, and tailoring chores are often too hard — or too boring. Middle-aged women who used to find satisfaction working in early childhood education often have difficulty cooperating with younger, second-generation women, whose style of work with small children is more easygoing.
>
> Older women frequently feel that their kibbutz career has not provided them with a skill, that women are more restricted in their choice of work than are men. Whatever the reasons, kibbutz women are less active than men in fulfilling prestigious tasks, such as the central-managerial ones, and they are less vocal in the weekly general meeting, where many kibbutz problems are decided. Quite a few women in the kibbutz still struggle with traditional feelings of female inferiority or dependence on male esteem. For most women in the kibbutz, then, it is not their work and social activity but their marriage and family that form the center of their lives.
>
> Women have often become proponents of *familistic tendencies* in the kibbutz. This term...denotes the demand of the family for greater authority in decision-making involving a member of the extended family, a demand frowned upon in kibbutz practice. It also conveys the family's desire to increase contact between parents and children by having the children sleep in the parent's apartment rather than in the children's houses.
>
> ...Supporters see the tendencies as a way to win back women who feel estranged from kibbutz life, but I find them regressive from two points of view. [The author then conveys his reasons against "familistic tendencies."]
>
> But the emergence of women's problems in the kibbutz raises nagging questions....If changed social conditions do not

bring far-reaching change in feminine characteristics, does that not prove the existence of an essential feminine character, rooted in biological structure? The scientific approach does not permit us to shy away from facts, even if they challenge our beliefs. But acceptance of this traditional image of women would mean renunciation of the equalitarian character of kibbutz society, and would entail a serious setback for women's emancipation movements everywhere. Before drawing conclusions, then, it would be well to examine the historical conditions that have affected kibbutz women. [The author goes on to analyze the kibbutz's stages of development, and then makes some conclusions.]

Thus, with all of the achievements of the kibbutz, two basic problems of women remain: Dissatisfaction in the sphere of work, and comparatively little participation in civic activities and the management of the society.

It would be easy enough to play down the problems I have raised. One could say that such dissatisfactions and tensions are typical of middle-aged women who are not interested in civic activity and careers, stop forcing them into a role that fits only your own utopian ideals of kibbutz society — and all the so-called problems will disappear! Perhaps. But other things will disappear as well, including the hope of active women fighting in the kibbutz and elsewhere for a change in the traditional image of women. And without this hope the kibbutz is doomed. Its very existence, as a socialist cell within a capitalist society, is a miracle. But if its women continue to find life frustrating, it is hard to expect the kibbutz to survive. [54]

This latter quote is from an article by Menachem Gerson, head of the Institute of Research on Kibbutz Education, Oranim, Israel, published in July, 1971, called "Women in the Kibbutz." His words are important to us for they were written by one who believes in the type of "equality" the kibbutz is striving for, but has not really obtained. In the great experimental society of the kibbutz, biology has raised its power and has prevented the "equality" radical feminists are pushing on us. In this quote by Gerson it should be noted that he makes value judgments about what kind of work is prestigious. *He in effect is saying that what is traditional women's work has little prestige and for women to get prestige they must do what men do or have traditionally done.*

3. Sex Roles

Avraham C. Ben-Yosef also has written on the problem of women in the kibbutz in his book, *The Purest Democracy in the World*:

> "The fact remains that, in general, the kibbutzim suffer from a shortage of women which, of course, affects the kibbutz social structure and, in the most serious cases, even its stability.
>
> There is a good deal of substance to the belief that it is more difficult for a woman than for a man to find complete satisfaction in the kibbutz, unless, of course, her personal relationships within the kibbutz are ideal for her. The belief is growing that the private housewife actually enjoys her almost constant work and worry entailed by her taking care of her house, husband, and children. Sometimes she frankly admits that this is the case."[55]

Kibbutz's Gender Equality

An internet article put it this way:

> In the first couple decades of the kibbutz there was not traditional marriage. If a man and woman wanted to get married, they went to the housing office and requested a room together. Not having traditional marriage was seen as a way to dissolve the patriarchy and give women their own standing without depending on a man (economically or socially) and was also viewed as a positive thing for the community as a whole, as communal life was the main aspect of the kibbutz.
>
> When the first children were born at the kibbutz, the founders were worried that this would tie the women to domestic service. They thought that the only difference between a man and a woman was that women gave birth and thus were automatically tied to the children and domestic duties. The communal dining and laundry were already a part of the kibbutz from the start. Of course they were implemented for reasons of living communally, but also to emancipate women from these duties so they were free to work in other sectors. With the arrival of the children, it was decided that they would be raised communally and sleep communally to free women to work in other fields. The desire to liberate women from traditional maternal duties was an ideological underpinning of the children's society

system. Women were "emancipated from the yoke of domestic service" in that their children were taken care of, and the laundry and cooking was done communally.

Interestingly, women born on kibbutzim were much less reluctant to perform traditional female roles. Eventually most women gravitated towards the service sector. The second generation of women who were born on the kibbutz eventually got rid of the children's houses and the Societies of Children. Most found that although they had a positive experience growing up in the children's house, wanted their own children at home with them.[69]

The documentary, Full Circle, summarizes the change in the women's view of equality on the kibbutz. The original Utopian goal of the founders was complete gender equality. Children lived in the children's houses. Freed from domestic duties, women participated in the industrial, agricultural and economic sectors alongside men. However, in the 1960s, while the rest of the Western world demanded equality of the sexes and embraced feminism, the second generation of kibbutz born women began to return to more traditional gender roles. They rejected the ideal achieved by their grandparents and returned to domestic duties such as cooking, cleaning and taking care of children. Today, most women do not participate in the economic and industrial sectors of the kibbutz. They even embraced traditional marriage.[70]

To Summarize. Much of this so-called problem of women in the kibbutz is that these women are in the wrong environment. Motherhood has been and shall be the prime profession of women as long as the human race continues to reproduce. Any society that denies this and fails to give due *value* to motherhood will have unhappy and dissatisfied women even though other factors such as food and material goods are abundant. The "equality" of the kibbutz is a failure as much literature confirms in more detail than what we have gone into here.[56]

---

[69] Paul Rothman dir. *Full Circle: The Ideal of a Sexually Egalitarian Society on the Kibbutz*. 1995. Filmmakers Library, 1995. Videocassette.

[70] https://en.wikipedia.org/wiki/Kibbutz as accessed on June 9, 2014

Although Margaret Mead's work was superficial and biased, and all experimental cultures have failed, another work by John Money and his associates seemed to be far more scientific and *seemed* to manifest a sex role neutrality-at-birth phenomenon.

Sex Role Neutrality-at-Birth Theory
Beginning of the Transsexual Myth

John Money. In the past, some knowledgeable radical feminists may have pointed out the studies of John Money, Joan and John Hampson of the Johns Hopkins Hospital which has since been discredited, but is still being pushed on society by the trans' cult of the 2020s.[71] These professionals believed in a psycho-sexual neutrality-at-birth theory. The psychosexuality of a person is his gender identity. In their dealings with hermaphrodites Money and the Hampsons came to the conclusion:

> "We conclude that an individual's gender role and orientation as boy or girl, man or woman, does not have an innate, performed instinctive basis as some theorists have maintained. Instead the evidence supports the view that psychologic sex is undifferentiated at birth — a sexual neutrality one might say — and that the individual becomes psychologically differentiated as masculine or feminine in the course of many experiences of growing up."[57]

According to them, males have no innate tendencies to behave in masculine ways after birth. They are born males, but not with masculine identity and masculine role behavior. They could just as easily be taught to behave in a feminine way, or to fit the feminine role even though they are born as males. That is, males can be reared just as easily to act and behave like females as they can be reared to act like males. Margaret Mead also believed that the sexes were this malleable. And radical feminists would love it, if it were true, for then they could bring up their daughters to behave just like men. Let's examine the psychosexual neutrality theory.

We'll examine two well-written critiques against this theory. Corinne Hutt of Oxford wrote a refutation of the theory in her book, *Males and Females*,[58] and Milton Diamond wrote a comprehensive

---

[71] as you will see in this book their theory turned out to be flawed

paper on it.[59] We will use some of their ideas and add some of our own.

Hermaphrodites. The whole idea of males and females being neutral-at-birth in regard to which role they will eventually play in adulthood, gives great emphasis to environmental and cultural factors. The idea implies considerably more malleability in infants than reality has heretofore manifested in mankind. Money and the Hampsons based their neutrality-at-birth theory on their studies of hermaphroditism. According to them a hermaphrodite is: "an individual in whom there exists a contradiction between the predominant external genital appearance on the one hand, and the sex chromatic pattern, gonads, hormones, or internal reproductive structures, either singly or in combination, on the other."[60]

Another definition of hermaphroditism by Money is: "As ordinarily defined, hermaphroditism or intersexuality in human beings is a condition of prenatal origin in which...the reproductive system fails to reach completion as either entirely female or entirely male."[61]

A hermaphrodite individual is sexually unfinished or partly male and partly female. It was with studies of such children that these doctors deduced their theory as far back as 1955.[62] But the infants studied were not normal males or females. They were intersexed. They were not males or females. They were hermaphrodites. We would expect such intersexed children to be more flexible in their gender role potential. To compare such atypical children with typical males and females is not the best proof, if it is any proof at all. Even when we closely examine Money and the Hampsons best arguments regarding hermaphrodites, we find much to be desired. They fall short in proving their thesis. In fact, in Money's 1972 book, *Man & Woman, Boy & Girl*, he seems to have conceded that humans are biologically biased at birth in some respects to either a male or female role direction because of the prenatal hormonal actions.[63] Money's change in attitude from his 1955 stand is due to the sound evidence that males and females have different organizations of the brain.[64] Nevertheless, Money still asserts in his book the greater importance of *post*natal experience: "much that pertains to human gender-identity differentiation remains to be accomplished after birth."[65]

In Money's book, co-authored with Anke Ehrhardt, it mentions that Chapter 7 and 8 are of possible use to the women's liberation movement. This is so because these chapters emphasize environmental factors. Although Money and Ehrhardt make the case for the interaction of biological and environmental factors as the explanation of behavior, they still emphasize, we believe unwarrantedly, environmental factors. They think there is a large potential flexibility in sex role behavior because of the alleged environmental factors in sex role development. Thus, because the book and papers of Money and the Hampsons are misused, we shall examine their psychosexual neutrality-at-birth theory even though Money has somewhat conceded the importance of biology's influence on sex behavior in his 1972 book.

Sex Role Assignment

The psychosexual neutrality-at-birth theory tries to prove that the gender role that one has been assigned by their parents at birth will remain his gender identity in adulthood. Money and the Hampsons in several papers have listed patients that were reared in a sex role *opposite* to their sex chromosome type (XY or XX), or *opposite* to their gonadal sex (testes or ovaries), or *opposite* to their hormonal sex (ratio of androgens to estrogens), or *opposite* to internal sex organs (Wolffian or Mullerian duct system), or *opposite* to external genital appearance (penis-like or vulva-like).[66] Almost all of the approximately 113 patients studied, with the exception of about 5 individuals, were said to have accepted their sex assignment and acted in accordance with it. Gender role reassignment to the opposite sex seemed possible in some cases, when it was done early enough. In order for gender role assignment to be effective the parents must not be ambivalent toward the assigned sex of their child (must not show doubts), they must assign the child's sex as early as possible (preferably before 18 months), and the child must also believe that he or she is of the assigned sex. If the child has doubts, then he will not accept the assignment. Sex reassignment after 18 months to 2 years is not advised.[67] According to Money and his co-workers, the fact that some hermaphrodite children were assigned and reared "successfully" as a member of a sex opposite to their gonadal, or their genital appearance is supposed to be proof that parents can teach any child to act out successfully either a male or female role. And this is because sex identity and behavior is neutral at birth, *according to the theory*.

Problems with Sex Role Neutrality-at-Birth Theory

(1) Hermaphrodites are not typical males or females — they are neither male or female; they are intersexual. Therefore they may seem more biologically malleable in regard to sex identity than normal males or females.

(2) Even though the world has many different ideologies, and deviations of sex play, the vast majority of the earth's people are reared in fixed traditional sex roles. If the human race is as old as some think, why haven't sex roles other than our traditional ones appeared more often than has been reported? If there is gender role neutrality-at-birth, where is the mass of cultural evidence of its existence in the form of contrary sex roles among large groups of people or among nations?

(3) At birth, infants are assigned their sex by appearance of their external genitals. Hermaphrodites are likewise assigned. In Chapter 4 we see the appearance of the external genitals indicates the influence of androgens on the hermaphrodite child (the more the androgenic influence, the more a penis-like organ appears). Contrariwise, the less the influence of androgens, the less the genitals look like a male's. When these children were assigned, they were assigned more as to what the child was, than what the child wasn't. In other words, the more the prenatal influence of androgens on a child, the more the chance the child's brain would be male-like, the more the chance his genitals would be male-like, the more the chance he would be assigned as a male, and the better the chance he will be effective in his role. The less the prenatal influence of androgens, the more the child would be female-like. Thus the more chance she would be assigned as a female, and the better her chance to be effective in such a role.

(4) Merely because the hermaphrodites don't outwardly seem to show the desire to give up their assigned roles, merely because they are erotically attached to their opposite sex, merely because they dress like their sex, merely because they perform their role, doesn't mean that they are as at ease with their role, or function as well in their role, as a typical male or a typical female. For example, those who were assigned as a female, but who were masculinized prenatally because their mother took hormonal injections in pregnancy for some ailment, acted male-like.[68] Money called them tomboyish. These "females" showed: (a) more athletic interest than normal females; (b) more self-assertiveness than typical females; (c) less self-adornment than typical girls in clothing, hairstyle,

cosmetics, and jewelry; (d) less rehearsal of maternalism in childhood, less enthusiasm for baby-sitting; (e) less interest in marriage and romance than interest in career and "achievement;" and (f) manifested visual erotic perception like males. Thus, Money's androgenized females performed their sex roles in a masculinized manner. If they were assigned as males, they would have performed more typically than they performed in their female-assigned roles.

(5) Hormones did influence the behavior of Money's hermaphrodites: (a) the prenatally androgenized females, whose mothers during pregnancy took masculinizing hormones, behaved in a male-like manner even though they were reared as females; (b) the hermaphrodites with the Turner's syndrome (XO chromosomes), who are *not* influenced by androgens in their prenatal state, were found to be "hyperfemales." That is, all the behavior typically known as feminine was abundant in these individuals. By comparing these two groups, one sees the influence of prenatal androgens (or absence of) on the infant's subsequent behavior.[69]

(6) Money and his associates used hormonal replacement therapy, cortisone therapy, and plastic surgery to correct hormonal levels and to make these individuals appear the same as those of their assigned sex.[70] Although these individuals may have begun with ambivalent and contrary hormonal influences and outward appearances, they were medically treated so as to be biologically like typical males or females. *This point, in itself, rules out the conclusions of Money and the Hampsons.*

(7) In the cases of an individual being reared in a sex role opposite to the sex appearance of his or her genitals, there were enormous problems "to surmount in coming to terms psychologically with their paradoxical appearance. It has been our experience that more than anything else, the *visible* anatomic genital or bodily contradictions occasion the greatest psychologic distress."[71] Even though these children's genitals were not as developed as typical individuals, they nevertheless suffered because of their contrariness. Moreover, it must be remembered, it was not only the contrary visual appearance, but also the contrary hormonal internal influence that made these individuals suffer. Normal males and females would find it even more difficult to overcome their biology. This difficulty encountered when an individual attempts to behave contrarily to his biological nature is another proof that normal persons are *not* neutral-at-birth in regard to gender identity.

(8) One other important proof against the neutrality-at-birth theory is that there are many cases of sex reversals after the so-

called critical development period. The critical development period for sex role development is between birth to 2 or 3 years of age, according to the theory. The theory says that up to 2 or 3 years of age, a child can be conditioned to behave and identify as either a male or a female. But after this critical period, the child finds it almost impossible to change his or her sex role. Although the neutrality-at-birth theory indicates great flexibility in role identity in infancy, it paradoxically says that after the critical period sex role identity is not changeable. But below we shall give some examples of sex reversals after the critical period. It should be noted that many of these changes occurred because the individuals did not feel at ease with their assigned sex roles. This uneasiness is probably due to internal biological pressures of the individual that are opposite to his or her assigned sex.

Dewhurst reported 20 cases of sex reassignment after 3½ years of age. Most were "successful," four cases were doubtful.[72] These children were brought up in one sex role, but in the assigned sex role, they manifested the opposite sex's behavior, and many wanted to be the opposite sex:

> "We believe (and this is all we claim) that, making due allowance for the difficulties we have mentioned, the results show that some children *can* have their sex changed after the age of 1½ to 2 years without disastrous results and perhaps with complete success.
>
> "Our records also provide some interesting information on the view of Money et al., that the sex of rearing is of such paramount importance in establishing the gender role. Although we agree that the sex of rearing is very important in this respect, some of these cases do suggest that the children had an affinity to the sex opposite to that in which they were being brought up."[72]

The authors note that just because there are cases where individuals have changed their sex role, doesn't mean it is easy to reassign one's sex role. It just means in some cases it can be done when it is to the advantage of the patient because he has biological and cultural tendencies toward the new sex role.

Berg reported a successful sex reassignment at puberty.[73] Diamond in his critique of the neutrality theory lists several other cases.[74] Diamond lists one particular case of interest where "an unambiguous male was raised from birth as a female." If gender role is neutral-at-birth, then such a child should behave like a female, but "despite attempts by the parents to make this child a girl, almost from birth on the child refused to be comfortable in the assigned sex or sex of rearing, continuously fighting all attempts from her feminine mother to be a feminine daughter." (p. 154)

The above points (1) to (8), are good evidence against the neutrality-at-birth theory. We will now turn to other so-called evidence of the neutrality-at-birth theory and of the socialization of sex roles. This supposed evidence is found in Money's and Ehrhardt's 1972 book, *Man & Woman, Boy & Girl*.[75]

My well worn copy in which I bought new in May 1973

Chapter 7 & 8 of Money's Book

Let's examine Chapter 7 of Money and Ehrhardt's book. The contents of both Chapter 7 and 8 were used by radical feminists.[72] (I first read and studied this book in 1973.) The evidence against the cases presented in chapter 8 of their book is much like the eight points just listed above. For one thing, the individuals named in Chapter 8 were given hormonal therapy and plastic surgery. Thus, they were not only reared in a certain sex role, but hormonally and appearance-wise prepared for the assigned role.

Two Cases

In Chapter 7, two cases were presented that were supposed to prove the "extraordinary influence" in shaping a child's sex role behavior of the parents' differentiated patterns in rearing the child.

First Case: *Identical Twin Boys*. The first case involved identical twin boys. One of them in his seventh month had his *penis burned off* in a surgical mishap during a circumcision operation. At the seventeenth month the parents finally decided to rear their son as a girl because of various medical advice from John Money. The twin boy was thus sex reassigned as a girl.

She (the reassigned twin boy) was then conditioned to behave as a girl. She was given girl's clothes and toys, given girl's tasks, and treated as a girl by the family thereafter.

But even though she was conditioned to behave like a girl she "had many tomboyish traits, such as abundant physical energy, a high level of activity, stubbornness, and being often the dominant one in a girl's group." (p. 122) Even though "her mother had tried to modify her tomboyishness" and even though the mother was very direct in conditioning this child to behave like a girl, this ex-boy, nevertheless, manifested typical boyish activity levels. In fact, she was "the dominant twin," she dominated the other boy twin like a "mother hen."

In a 1983 book by Jo Durden-Smith and Diane de-Simone, *Sex and the Brain*, they wrote about this case:

> Set against all this, of course, is still the one case of the American male identical twin, surgically altered soon after

---

[72] Although some of the evidence from Money and Ehrhardt was questioned, gender radicals still persist in preaching versions of the neutral-at-birth theory.

birth and raised successfully, all accounts, as his brother's sister. "You have to understand," says Milton Diamond of the University of Hawaii, "that this one case was seen, and is still seen, as being of absolutely crucial theoretical importance. Throughout the 1970's, it was written up in an enormous number of textbooks in psychology and sociology. It was included in virtually *every* book on sex differences, every book that addressed itself to the roles, in this society, of men and women."

In 1965, Milton Diamond wrote one of the first scientific papers that attempted to gather together all the existing evidence that might support the idea of the prenatal sexual differentiation of the human brain. And in following years, in a series of other papers, he went on to buttress his case. In 1979, then, he was approached by the **British Broadcasting Corporation** [my emphasis] for his help in a film the producers wanted to make about the American twin. They had already talked with John Money and had secured his assistance — he was to be the leading voice in the program. And Milton Diamond, because of the view he was known for, was to be a sort of foil.

"Well, the producers went off to do their filming," he says. "And they talked to a number of psychiatrists who'd been first introduced to the child when she was about thirteen, some three years before. It was plain that the child had *not* made the successful gender switch that has been claimed for her. She was having major problems. She'd refused to talk about any difficulties she'd had in the past, and had been reluctant to talk about sexual matters at all. She had shown considerable ambivalence about her position as a female.... She was feeling that boys had a better and easier life and wanted to be a mechanic. She looked quite masculine. And she was described as unhappy and ambivalent about her status. One psychiatrist said: 'She is having considerable difficulty in adjusting as a female. At the present time she does display certain features which make me suspicious that she will ever make the adjustment as a woman.'"

When the BBC told John Money what they had found, he simply withdrew his support and refused to be interviewed. The program, however, was aired in Britain at the beginning of 1980 without him. And since then Money has failed to address the issue in print, though his version of the case's outcome is still everywhere quoted.

"But it [cases like the twin boys case] *certainly* doesn't support," he says, "the idea that gender identity and sexual orientation are dependent entirely on social *learning* — which is the one and only idea being persistently peddled in the sort of books I was talking about. This is something that everyone — scientists included — ought to face, and face squarely. All the evidence so far gathered points to the fact that the foundation — the fundamental directionality — of a man or a woman's future sexual identity is laid down in the masculine or feminine brain *before* birth."[73]

[A transcript of the 1980 BBC program can be found at https://www.bbc.co.uk/sn/tvradio/programmes/horizon/dr_money_trans.shtml The video of the program has been taken off Youtube... I wonder why.]

The **New York Times**, on their front page in the spring of 1997, March 14, had an update to this story.[74] Despite everyone telling him constantly that he was a girl and despite his being treated with female hormones, his brain knew he was a male. Eventually his father told him the truth and he had surgical help to repair his sexual organs enough so he married and had sexual relations as a man. He eventually committed suicide in 2004.[75] But this one case more than any other was used by the gender radical to push their neutral-at-birth theory. Their theory is dead wrong and has been misused to further confuse the sexes.

Second Case. The second case given in Chapter 7 of Money's book was about a genetic male (XY chromosomes) with a very small penis. "The phallus was 1cm long, so small as to resemble a slightly enlarged clitoris, and like a clitoris, it did not carry a urinary canal." (p. 123) Although a female sex assignment was suggested to the parents at birth, the parents on the advice of a specialist decided to

---

[73] Jo Durden-Smith, JO and Diane de Simone, *Sex and the Brain*, Warner Books:New York, 1984 (1983), pp. 74-76.

[74] https://www.nytimes.com/1997/03/14/us/sexual-identity-not-pliable-after-all-report-says.html time machine: https://timesmachine.nytimes.com/timesmachine/1997/03/14/issue.html

[75] https://en.wikipedia.org/wiki/David_Reimer and https://www.youtube.com/watch?v=0Zw1EdRKocI&t=20s

rear their child as a boy, "despite the absence of a penis." But after many months of weighing their decision, the parents decided at the seventeenth month to reassign the child as a girl. Because the so-called penis of this child was more like a clitoris, and because the child was so young this reassignment was apparently "successful," according to Money, through parental conditioning of the child. However, this child manifested tomboyish behavior, and "would say occasionally that she was a boy and not a girl." (pp. 124-125) When she was three, her "behavior still seemed quite tomboyish...she also still seemed to have more physical energy expenditure." (p. 125)

These two cases were supposed to show that parental rearing has "extraordinary influence on shaping a child's psychosexual differentiation and the ultimate outcome of a female or male gender identity." (pp. 144-145) **We disagree.** The evidence we have presented is contradictory to their neutrality thesis. Events after birth do have a great effect on children, but these events do not overcome or suppress the children's biological underlying sex differences. The prenatal hormonal influences caused these two "girls" described in Chapter 7 of Money's book to behave in rough tomboyish ways. This was the case even though they were conditioned by their parents to be females. Furthermore, these children were to receive hormonal treatment as they developed. This would enable them to develop in their assigned sex role properly only **if** Money's theory was correct. But it wasn't.

In the last part of Chapter 7 Money takes a look at cross-culture studies which seem to indicate flexibility in sexual partnerships of humans. Of course, it is possible for otherwise typical individuals to perform homosexual acts or bisexual acts. But these possible variations of sexual acts among mankind are not proof of the overwhelmingness of environmental factors as Money's work tries to indicate. Without the majority of a society practicing heterosexuality, the society would not last long. There never has been nor will there be a homosexual nation or state. It is biology that predestinates mankind to heterosexuality, not social conditioning, as we will see.

Although Money presents several cases of flexibility in sex play, nevertheless, in his presentation we frequently see manifestations of the underlying biological sex differences. Among the Melanesian people, "boys are made fun of for having an erect penis." Yet, "they continue to play with it, nevertheless." (p. 136) This shows young males typically more asocial behavior.

Contrariwise, girls are scolded "for touching their genitalia in public. Soon they cease to do so." (p. 136) Unlike the males, girls show prosocial behavior as typical females frequently do. Both sexes were taught not to play with their genitals, both reacted differently.

Sex Role Development and Environmental Influences

In the paper, "Sex Roles and the Socialization Process,"[76] the author, Sverre Brun-Gulbrandsen, tried to determine "the extent to which actual differences in behavior and attitudes between the sexes can be traced to environmental influences....We do not exclude the possibility that biological differences can be significant in the cases we study; we merely focus our attention on environmental variations." (p. 62) Although the author was aware that sex differences "may stem from biological differences" he chose to concentrate on environmental variations much like the radical feminists.

The radical women liberators emphasize and re-emphasize environmental causes of behavior, while relating biological differences as secondary. To them biology is no more important in understanding sex behavior than sex-differentiated clothing.[77] Therefore, since Brun-Gulbrandsen emphasized environmental factors like the radical feminists, and since he wrote his paper well, we shall take a look at his paper.

The main conclusions from Brun-Gulbrandsen's paper about sex role development were:

1. Even boys and girls, who were brought up "by mothers who profess verbally to believe in equality and similarity in childrearing," have the rules and norms of traditional sex roles "well internalized." (p. 66-67)
2. "Because the parents have so completely internalized their views of social sex roles they are unaware that they clearly and systematically influence their sons and daughters through numerous negative and positive sanctions to accept different behavior patterns." (p. 67)
3. "Children are not led to accept the sex role pattern by force or violence, but usually by more subtle means." (p. 69)

    (a) First they are encouraged to do a typically masculine (or feminine) activity and then rewarded for doing so. This social reward makes such activity enjoyable, and so he (or she) continues in such behavior.

(b) Another means is to use negative verbal reinforcement when he or she does something not appropriate to his or her roles.

(c) All these small "pushes" in "systematically different directions" make the resulted sex roles seem natural.

4. A major obstacle to sex equality is that "people interpret an attempt to introduce greater similarity to sex roles as an attempt to change human nature." (p. 70)

5. As socialization of sex roles change and become more equal, what was once thought of as natural, will show itself to be "a product of early indoctrination."

6. But because the traditional system has gained such a firm hold "parents consciously [among those who believe in traditional sex roles] or unconsciously [among those who do not believe in traditional sex roles] rear their children largely in the same way as they were reared: change is introduced, but slowly and over a span of generations." (p. 77)

Parenthetically, this last conclusion (6) is probably why Simone de Beauvoir quoted Stendhol, "The forest must be planted all at once," for she understood that perfect equality could only come according to the socialization theory if all social pressures were negated at once. Otherwise, equality as radical feminists describe it will only come *slowly*, if at all.

The findings (1) to (6) sum up in a few words how radical feminists understand and relate the socialization of sex roles. These points are environmental ideas on how sex roles have come about. But these points do not answer the question: why are sex roles almost universally the same in all cultures? When we say "sex roles" here, we mean the traditional ones: most men work outside the homes, work at more physically strenuous and riskier tasks, dominate all spatial ability jobs (mathematicians, engineers, architects), run the major ruling institutions, etc; women work in and around the homes most of the time, care for the children, cook, sew, etc., and when they work outside the home they work in more occupations that deal with children and interrelationships between people and occupations where verbal skills are important.

Brun-Gulbrandsen also asks the same question about why the traditional gender roles are so universal: "But why is it that certain forms of behavior are ascribed to the male role and others to the female role? This interesting problem has yet to be solved...." (p. 72) The reason this "interesting problem" hasn't been solved is because the author merely focuses his attention on environmental variations instead of bringing the biological variables into the picture. It is not scientific to leave out the biological variables, but Brun-Gulbrandsen did and radical feminists do.

Brun-Gulbrandsen writes about children of both sexes aged 8, 11, 14, and 15 years, who were asked to identify various kinds of behavior, as either being what boys usually do or what girls usually do. The great majority of these children identified antisocial and delinquent behavior as being the behavior of boys. The children felt that girls were more pro-social. Girls in the eyes of the children seemed to possess qualities of behavior deemed good by society. Brun-Gulbrandsen then asks, "One may wonder whether it is possible for an ordinary 8 year-old to have such a clear image of asociality in male nature." He then answers his question, "The stereotypes are 'internalized'."

They are internalized, according to the author, through the children's continued subjection to the "notion" that males are asocial. And because of our society's view, males are "permitted" to show more aggression. To the contrary, females are not allowed much show of aggression. If they show aggression, various negative labels are thrown at them. Walter Mischel describes a similar process of socialization in his paper in the book, *The Development of Sex Differences*.[78] Therefore, sex roles or gender roles are primarily learned, at least according to the radicals' way of thinking. Nothing could better illustrate radical feminists' thinking on this than this quote from a feminist: "I think it is the consensus that men are sperm donators, women are baby incubators, and all the rest of it is the result of the socialization process."[79] And this same feminist thinks, "The Women's Liberation Movement is our [radical feminists] force to break through the chains of socialization." (p. 492)

Asocial & Prosocial Behavior

Let's examine this asocial and prosocial behavior of boys and girls respectively. Brun-Gulbrandsen,[76] says that males are more asocial merely because they are permitted to behave that way; girls are not allowed to show aggression and are taught to be nice young ladies. Walter Mischel reported similarly,[80] as do many of those pushing women's "liberation." But Mischel also said: "Unfortunately [for the environmentalists], present evidence that the sexes are indeed treated differently by their parents with respect to the above behaviors [dependence & aggression] is far from firm...." (p. 75) When researchers who believe in the learning theory find parents with equalitarian ideas on childrearing, but that have children with obvious sex differences in behavior even though they reared them to be sex-role neutral, they are forced to rationalize and explain away their findings. They point to some dubious "unconscious attitudes" in the parents that somehow influence the children's behavior.[81]

Let's examine a study that revealed obvious biological influence on sex behavior, but the authors nixed it for some extreme rationalization because the contrary results were not the desired results wanted by the socialization authors.

In a study by Sears, Maccoby, and Levin, they interviewed 379 mothers of children in public-school kindergarten in the Boston area. They were looking for differences in child treatment by the mothers that might help to explain the differences in gender role behavior of children. In Sears' words:

> "The results of this inquiry are astonishing and disappointing. In the first place, almost no relationships were found, as indicated by correlations, between childrearing practices and antisocial aggression, in either sex. The same was true for prosocial aggression in boys. But the astonishing part of the results comes with respect to the girls, and the antecedents of their prosocial aggression....We find a clear picture of a significant relationship between high use of *masculine* child-rearing procedures [physical punishment]...and high *prosocial* aggression...The results are exactly the opposite of what one would expect, namely, that feminine (prosocial) aggression would be greater in those girls who were more femininity treated [with withdrawal of love method, or verbal punishment]...."[82]

After stating the results, the authors move on and intellectualize away the contrary findings. But we can see the results as clearly helping to prove that biological factors do influence the sex behavior found in the study. Boys showed more asocial aggression in the study because of the androgenic influence: (1) that makes their bodies more active than girls; (2) that makes their bodies in need of more physical activity than girls; and (3) that causes them to express their aggressive behavior in more physical ways (which seems more destructive than girls' more refined aggression). Girls to the contrary, showed more prosocial aggression due to the physical punishment from their parents because: (1) girls show more sensitivity to discipline than boys (they don't want to be spanked again, therefore they show their aggression in more indirect ways); and (2) they have a greater innate need for parental approval than boys. This prosocial behavior by girls is *not* because they learned it through differentiated parental discipline.

The question is, should we change our society for the views of the radical feminists? Some of their ideas have been activated in our society over the last 40 years but have only made more unhappier women. There have been social experiments with the "equality" that the women liberators are propagating. But they have failed and will continue to fail. The main sexual behavior differences between men and women have to do with biology, not the environment, not socialization.

Functional Differences and Behavior

From a study of cross-culture papers by Herbert Barry and others, the following is pertinent to our discussion in this chapter:

> The childbearing which is biologically assigned to women, and the child care which is socially assigned primarily to them, lead to nurturant behavior and often call for a more continuous responsibility than do the tasks carried out by men. Most of these distinctions in adult role are not inevitable, but the biological differences between the sexes strongly predispose the distinction of role, if made, to be in a uniform direction.
>
> The relevant biological sex differences are conspicuous in adulthood but generally not in childhood. If each generation were left entirely to its own devices, therefore, without even an older generation to copy, sex differences in role would presumably be almost absent in childhood and would have to be developed after puberty at the expense of considerable relearning on the part of one or both sexes. Hence, a pattern of child training which foreshadows adult differences can serve the useful function of minimizing what Benedict[83] termed "discontinuities in cultural conditioning."
>
> The differences in socialization between the sexes in our society, then, are no arbitrary custom of our society, but a

> very widespread adaptation of culture to the biological substratum of human life.[84]

This important quote gives us the real reason we rear our children to be sex-differentiated in certain ways.

In the book Family, Socialization and Interaction Process, by Parsons and Bales,[84a] we read:

> In our opinion the fundamental explanation of the allocation of the roles between biological sexes lies in fact that the bearing and early nursing of children establish a strong presumptive primacy of the relation of mother to the small child and this in turn establishes a presumption that the man, who is exempted from these biological functions, should specialize in the alternative instrumental direction.(p. 23)

We agree with this position, and an unbiased analysis of the human situation manifests this clearly. Only the most radical are blind to nature and its calling.

Women Have Babies; Men Do Not Have Babies

Women have babies; men do not. This is the main (but not the only) cause of the differences between sex roles. Mothers have closer biological ties with their children than fathers because of the nine months of pregnancy and its hormonal effects. Judith M. Bardwick, a professor in psychology, in her book, *Psychology of Women*, has put it this way:

> I feel that there is a biological origin of maternal nurturance, especially of small and helpless infants. My reasons are intuitive, unscientific, and unsupported. A few years ago I watched an enormous colony of monkeys chasing one frantic female. The lone figure ran, terrorized, screaming and panicked, away from the chasing throng. In her arms she held the rigid corpse of a dead infant. In all the time I watched, the colony never succeeded in separating her from that body. I have been impressed with the exquisite tenderness older children, especially girls, show toward young children. My own daughters never played with dolls, never rehearsed maternal roles with toys. Their response to young children and animals is nurturant and gentle. I myself had never seen an infant before my eldest was born. Preoccupied with school and work, I had never given much thought to even my own imminent child. My response to my child was immediate,

> unrehearsed, and unexpected. In a second she was incredibly precious — to be loved and protected. My husband's response I remember better than my own: "No man will ever be good enough!" The anecdote is glib but my intent is serious. My reaction to my child was not comprehensible in terms of the differential reinforcements of learning theory. This was a primitive, gut-deep response, not amenable to easy verbalization, similar in kind and intensity to that of the monkey who could not give up the body of her dead infant. I have found that this feeling diminishes slowly as my children grow and successfully achieve their independence. But each new birth, each new child, brought forth the same immediate, profound surge of love and protectiveness.[85]

Bardwick goes on to state some proof that maternal tendencies are caused biologically through hormone levels and/or because the sexes, as some good data indicate, have sexually differentiated brains. Brizendine in her book, gives further credence to this in her book, *The Female Brain*:

> Motherhood changes you because it literally alters a woman's *brain* — structurally, functionally, and in many ways, irreversibly. ... Deeply buried in my genetic code were triggers for basic mothering behavior that were primed by the hormones of pregnancy, activated by childbirth, and reinforced by close, physical contact with my child.[76]

See Chapter 4 of this book and read Brizendine's book for further information on hormones and maternal behavior. Of course, these motherly tendencies can be turned off by adverse environmental conditions. A few mothers even come to hate their babies. But in most cases mothers love and care for their children because of the changes in their brain caused by the different hormones, especially oxytocin.

Even in the days when I wholeheartedly believed in some radical aspects of feminism and culture conditioning, I observed with some amazement the warm and spontaneous reaction of girl friends and other women when they were around infants. I have met very few men who come close to the warm responses that the typical woman manifests around infants and children.

---

[76] Louann Brizendine, M.D., *The Female Brain*, 2006, Broadway Books, Chapter 5, "The Mommy Brain."

Real Reasons for Sex Roles

Because of women's biologically close relationship to infants — they grow them, they nurse them — societies of all kinds in all ages have women in the important role of caring for children. Human children need much more caring for than any other beings on earth. The human child needs a great deal of teaching, training, and loving in order to survive and function in this world. Mothers have traditionally done the care-taking for the world's children. Therefore, because of the nature of children, homes were established. And because women were at home caring for the children, they also performed work *around the homes and in them*. Thus, women traveled less because of their jobs at home. This helped to set up a division of labor, where men performed the needed work away from the homes because women were busy with the important care-taking of children at home.

Furthermore, since women are relatively less muscular than men, men performed the heavier work, and a division of labor due to strength was created. In this paper, earlier, we related how this very same process happened in Israel's kibbutzim to a certain degree in the last century. Other relative physical and mental differences have also contributed to the divisions of labor between the sexes. (see Chapter 4)

Science to Help?

Now some radical feminists admit that the functional differences between the sexes have caused many of the behavioral differences, the sex roles, and the division of labor in the past. But they insist that the industrial revolution and science have the means to set women free from motherhood. They claim that artificial insemination, test-tube babies, child care centers, and machines will set women free to enter on equal terms with men in the labor market. They also mention that because of the population boom, there will be less need for mothers, and thus, this will set many women "free" from motherhood.[77] But this overlooks important factors like the following:

---

[77] In the mid 20th century there was a population increase yearly of 2% — a "population boom." In 2025 this boom has *decrease* to about .85%, and by the year 2080 it will start decreasing, if today's projects are valid. [Grok 2025 search]

1. Most people of the world live under conditions of little industrialization.[86]
2. Pollution, economic, and energy problems will keep total or indefinite growth of industrialization from becoming reality. [87]
3. The economics of most industrial nations are based on growth.[88]
4. Because of (3) any society that achieves a zero population growth rate must increase production to survive in this competitive world. Yet (2) rules this out in the long run. The fact that modern zero population growth societies (with good healthcare) have more older retired people (65 and older—e.g. Japan [30%], Germany [21%]) than a growing society, helps to rule out increasing production. Zero population growth leads to a greater percentage of older people, which has a drain on the economy.[78] This is another case for women having more babies than one or none.
5. Artificial insemination is a method most, if not all, men are psychologically against, if they are fertile. Artificial insemination in mass will never happen, for men rule all powerful institutions.
6. Test-tube babies (grown totally outside the woman's body) are nothing but a fictional notion.[89] There are many problems to be solved before this can ever take place. It may never be possible because of the huge complications science must overcome. But if it were possible, only through some mad leader in some future age could this be brought about in a mass form against mankind's psychological needs. Such babies would have no mothers who would truly care for them since there would not be any true and close biological mothers. Test-tube babies would be deprived babies, and deprived babies are inferior children, and make unstable and inferior adults.[90]
7. The idea of child care for everyone would not work either, because child care centers are inferior for rearing children. [91] Furthermore, they are expensive, and therefore

---

[78] "World getting 'super-aged' at scary speed." Aug 21, 2014. http://money.cnn.com/2014/08/21/news/economy/aging-countries-moodys/?hpt=ob_articleallcontentsidebar&iid=obnetwork

3. Sex Roles

economically infeasible for many. Also women mostly care for the children in day care centers, so women would not be freed by this method. They would only directly or indirectly be caring for *other* children besides their own, but in a much more disadvantageous setting for the children. The taxation for mass child care centers would force many mothers to work and thus their own children would be forced into these inferior institutions. (Of course in *some* family situations, like ones with too many children and too little money where the mother can't properly care for her children, day care centers *may* be better, but not as good as a proper mother-child relationship.[92])

8. Most women like the work of caring for their children; they merely work at jobs to supplement their husband's income, or because they have no husband.[93] And even career women in top jobs feel that the family comes first if a conflict should arise between work and family, as a Political and Economic Planning report called *Women in Top Jobs* indicated.[94]

Therefore, because of the reasons given above (1) to (8) and other reasons, most women in most countries will continue to be in the profession of motherhood. We also see a trap here, where population growth leads to an over-populated earth, but zero growth leads to an overly aged population, which leads to poorer economies. Wars in the past took care of the overpopulation problem, but wars and poor healthcare are not the answer in a humane world.

Working Mothers with Children

Radical feminists may point to statistics that tell us that more women today *with* children are working at full-time jobs outside the homes than previously.[95] According to a 2008 Labor Department study 71% of mothers with children under 18 were employed, while 62% of the mothers with children under 6 were employed. 2013 figures were approximately the same as the 2008 figures.[79] In 1960 this percentage of working mothers with children under 6 years of age was only about 18%. But this phenomenon does not mean that biology does not tend to dictate different and universal roles for each sex. For these mother's the dual roles are creating more stress, friction and less-adjusted children than would be otherwise. When

---

[79] Bureau of Labor Statistics — http://www.bls.gov/news.release/famee.nr0.htm

you go against biological tendencies you create problems, not solve problems. Some of the reasons more women are working today outside the home are:

1. birth control and abortion enable women to work for years without interruption because of unintended pregnancy;
2. fast food restaurants, canned foods, microwave ovens, and frozen foods grant women more freedom from long hours of food preparation; schools are in effect being used as child care and feeding centers by some parents so they can work;
3. their families can't buy a house or even rent a place unless they work because of the economic reality;
4. women are entrapped in material desires propagated by the media and the national spirit, and they feel (or their husbands feel) that they must work to support these desires;
5. women are indoctrinated by radical feminists and other male chauvinists to think they are only valuable when they work outside the home;
6. some of the tasks that should be theirs, like teaching their children to read and write, have been taken over by the state and this has therefore taking away some of their challenges and rewards of staying home.

Note: Some of these biological differences mentioned in the two sections below will be amplified on in the next chapter of this book.

What the Socialization Theory Cannot Explain

As we have tried to show herein, the socialization theory cannot explain many aspects of sex differences. The following are some of the facts that socialization cannot explain:

1. It can't explain why the majority of the world's people live within the traditional sex roles, and always have as far as records show.
2. It can't explain why sex-differentiation is in the same direction everywhere: most males act in traditionally masculine ways; most females act in traditionally feminine ways.

3. It can't explain the tomboyish behavior of the female androgenized hermaphrodites.
4. It can't explain the "hyper feminine" behavior of those with Turner's syndrome.
5. It can't explain the sex differences in spatial ability or verbal ability.
6. It can't explain the differences in the vigor of activity between boys and girls.
7. It can't explain boys' more aggressive, assertive behavior, or their more asocial behavior in comparison to girls.
8. It can't explain girls more affiliational needs in comparison to boys.
9. It can't explain physical differences between the sexes in such aspects as strength, height, maturational rate, and so forth.
10. It can't explain the differences between the male and female brains. In the next chapter we will document this assertion.

Conclusions on Sex Role Development

Sex roles are the way they are in this world because:

Biology limits and prepares each sex in different ways, which in the long run has produced the traditional gender roles we have today. The traditional sex roles cut across all cultures because biology has dictated limits to how far each sex can stray from their innate behavioral tendencies.

Biology limits and causes a bio-cultural pattern of behavior. Since males and females are biologically dissimilar everywhere in the world, their biology limits and causes a universal bio-cultural pattern among mankind.

This bio-cultural pattern is incorporated within the thoughts and institutions of mankind. The incorporation of behavioral patterns within the thoughts and institutions of mankind reinforces and magnifies the behavior patterns.

The parents reared in the universal bio-cultural pattern, teach their children, directly or indirectly, consciously or unconsciously, to behave according to their proper gender roles.

The children accept their assigned gender roles because their sex pertinent innate influences make it easy for them to accept their sex assigned roles.

If parents fail to rear their children in the traditional sex roles, for ideological reasons or from ignorance, then the children are hampered in their development by: (a) their opposing innate biological drives; (b) the opposing cultural pressures from those properly reared.

If a whole society tries to go against the historical reality of traditional sex roles, then that society helps to diminish itself because: (a) the society drives each individual against their innate biological drives (which weakens the individual); and (b) the society because of (a) begins to fall behind other societies in productivity/well-being since too much energy is used to promote their ideology and to fight against their biology.

In summary, we see that there is much less cultural or environmental dictation of sex roles than declared by radical feminists, and with evidence from the next chapter (as well as this

3. SEX ROLES

one) we see why biology is the main cause of behavior. Although it is true that cultural and other environmental factors have a part in affecting behavior, the interaction of biology and milieu is not the main cause of behavior. A horse acts different from a lion mainly because of its biological construction not because of environmental factors. Men and women are different because of biology, not because of culture.

Men and women think differently because they each perceive the world differently: there is a male brain; there is a female brain (see next chapter).

Now let's turn to sex/gender differences and their biological connection.

References for Chapter 3

[1: ch 3] is 109 of the Bibliography list

[2: ch 3] 95
[3: ch 3] 36; 85
[4: ch 3] 207
[5: ch 3] 49
[6: ch 3] 49
[7: ch 3] 86
[8: ch 3] p. 1451 in 111
[9: ch 3] p. 260 in 35
[10: ch 3] p. 261 in 35
[11: ch 3] p. 260 in 35
[12: ch 3] p. 259 in 35
[13: ch 3] pp. 259-260 in 35
[14: ch 3] p. 16 in 109
[15: ch 3] p. 293 in 109
[16: ch 3] 184
[17: ch 3] p. 53 in 110
[18: ch 3] p. 389, 392, 395 in 110
[19: ch 3] p. 246ff in 110
[20: ch 3] p. 377 in 110
[21: ch 3] p. 377 in 110
[22: ch 3] p. 387 in 110
[23: ch 3] p. 395 in 110
[24: ch 3] pp. 389-395 in 110

[25: ch 3] 20

[25a: ch3] p. 1455, in 111
[26: ch 3] pp. 321-322 in 138
[27: ch 3] pp. 395-396 in 110

[28: ch 3] **190**
[29: ch 3] p. 392 in **110**
[30: ch 3] p. 1438 in **111**
[31: ch 3] pp. 462-463 in **4**
pp. 158-160 in **151**
[32: ch 3] **91**
[33: ch 3] p. 111 in **109**
p. 1441 in **111**
[34: ch 3] p. 1439 in **111**
[35: ch 3] **160** & **161**
[36: ch 3] pp. 54-55, 114-115, 170, 175,
182, 186-191, 226-227, 234 in **109**
p. 1439 in **111**
[37: ch 3] pp. 49, 132-133, 173
183-184, 246 in **109**
pp. 322-323 in **138**
pp. 636-637 in **142**
[38: ch 3] pp. 389, 392, 396 in **110**
pp. 37-39, 54, 168, 180, 227, 237 in **109**

[39: ch 3] pp. 176ff in **38**
[40: ch 3] pp. 188ff in **38**
[41: ch 3] pp 252-255 in **109**

[42: ch 3] p. 1437, footnote in **111**
[43: ch 3] in **189**
[44: ch 3] p. 220 in **54**
[45: ch 3] pp. 225ff in **54**
[46: ch 3] pp. 221, 223 in **54**; in 1959 there were 30 million
    women working versus 18.6 million men working in
    Russia
pp. 133ff in **168**
[47: ch 3] pp. 227ff in **54**
chapter 5 in **168**
[48: ch 3] pp. 130, 134, 72 in **168**
p. 55 in **11**
[49: ch 3] **172**
[50: ch 3] p. 9 in **98**
[51: ch 3] pp. 128-129 in **98**
[52: ch 3] pp. 178 ff in **38**; **64**
pp. 80ff in **22**; **98**
[53: ch 3] pp. 221-230 in **173**
[54: ch 3] pp. 567, 568, 572 in **64**
[55: ch 3] p. 82 in **22**

3. SEX ROLES

[56: ch 3] <u>22</u>; <u>64</u>; <u>89</u>; <u>98</u>; <u>173</u>
[57: ch 3] <u>76</u>; see <u>122</u>
[58: ch 3] <u>86</u>
[59: ch 3] <u>45</u>
[60: ch 3] <u>76</u>
[61: ch 3] p. 5 in <u>120</u>
[62: ch 3] <u>122</u>
[63: ch 3] p. 1-2, chap. 6 in <u>120</u>
[64: ch 3] chapter 6 in <u>120</u>
[65: ch 3] p. 114 in <u>120</u>
[66: ch 3] <u>75</u>
[67: ch 3] <u>76</u>
[68: ch 3] <u>120</u>
[69: ch 3] **119**; pp. 111-114 in <u>120</u>
[70: ch 3] <u>120</u>; <u>76</u>
[71: ch 3] <u>75</u>
[72: ch 3] <u>44</u>
[73: ch 3] <u>24</u>
[74: ch 3] <u>45</u>
[75: ch 3] <u>120</u>
[76: ch 3] <u>30</u>
[77: ch 3] <u>109</u>
[78: ch 3] <u>100</u>
[79: ch 3] p. 493 in <u>196</u>
[80: ch 3] <u>112</u>
[81: ch 3] <u>188</u>
[82: ch 3] p. 145 in <u>156</u>
[83: ch 3] <u>19</u>
[84: ch 3] pp. 206-207 in <u>14</u>
[84a: ch 3] <u>138</u>
[85: ch 3] pp. 33-34 in <u>9</u>
[86: ch 3] chapter 35 in <u>192</u>
[87: ch 3] pp. 100-101 in <u>192</u>; <u>79</u>
[88: ch 3] pp. 100-101, 339-341 in <u>192</u>; <u>214</u>
[89: ch 3] <u>186</u>; <u>203</u>
[90: ch 3] <u>210</u>
[91: ch 3] <u>210</u>; <u>97</u>
[92: ch 3] <u>210</u>
[93: ch 3] p. 276 in <u>90</u>
[94: ch 3] <u>55</u>
[95: ch 3] <u>135</u> see latest almanac

4. Real Sex Differences and Their Implications

Clearly, when one reads the dogma of radical feminists, one is faced with their aversion, if not denial, of biologically caused sex or gender differences. They think that socialization causes most sex differences or gender differences. "We are socially conditioned to act in different ways," so they say. The following statement given before the House of Representatives subcommittee studying the Equal Rights Amendment to the United States Constitution says it the way the radical feminists believe and think:

> The Women's Liberation Movement is our force to break through the chains of socialization.... I think it is the consensus that men are sperm donators, women are baby incubators, and all the rest of it is the result of the socialization process.[1]

This belief that most sex differences are caused by socialization goes far back in intellectual circles. Amram Scheinfeld in his 1943 book, *Women and Men*,[80] mentioned a little about this:

> The objective [of his book] has remained the same: To help women and men toward a better understanding of themselves in relation to each other....According to the original outline, I had expected to devote myself mainly to the social factors... and to give only passing attention to biological sex differences. In this approach I was reflecting the prevailing tendency among social scientists to regard differences between women and men in behavior, thought, temperament, and achievement, as chiefly the products of "conditioning."
>
> But as intensive research proceeded... it began to appear that the original premise had many weaknesses. The basic sex differences, I was forced to conclude, were far more extensive. [p. ix]

Scheinfeld used the first chapter of his book to discuss why there was "the antagonism or indifference to the investigation and discussion of sex differences." (p. xi) One aspect, he felt, was that some used differences to support the domination of one sex over the other, or for:

---

[80] *Women and Men*, Amram Scheinfeld, Harcourt, Brace and Co, NY, 1943

maintaining the barriers between them, particularly to keep women "in their place" and to prevent individual women, regardless of their capacities or ambitions, from proving their worth in competition with men.

Little wonder, then, that the stressing or even the investigating of facts regarding basic sex differences became distinctly unpopular in liberal circles. [p. 6, *Women and Men*]

From the *Female Brain*,[81] Dr. Louann Brizendine writes:

> In the 1970s at the University of California, Berkeley, the buzzword among young women was "mandatory unisex," which meant that it was politically incorrect even to mention sex difference. [...] The fear of discrimination based on difference runs deep, and for many years assumptions about sex differences went scientifically unexamined for fear that women wouldn't be able to claim equality with men.

But as she says, this means "perpetuating the myth of the male norm" which "hurts women" and "ignores the different ways that they process thoughts" as well as "undervaluing the powerful, sex-specific strengths and talents of the female brain." In her book:

> In writing this book I have struggled with two voices in my head— one is the scientific truth, the other is political correctness. I have chosen to emphasize scientific truth over political correctness even though scientific truths may not always be welcome.

What is shocking to me is that in most scientific studies on sex differences they usually start out their paper or book apologizing for going against the socialization/nurture bias at their institutions. From this we see how well the radicals have done their job of indoctrination in our colleges.

In the previous chapter of this book we examined socialization and how much it does or does not do to create or enforce sex/gender differences. We started by examining Margaret Mead's work dating back to the 1930s. We saw how the old-radical feminists were mistaken in their belief about the overriding influence of socialization on sex/gender roles. They practically deny that any biological differences exist. But even at the beginning of the latest installment of feminism (1960s-70s genesis) there were many

---

[81] Louann Brizendine, M.D., *The Female Brain*, 2006, Broadway Books [Random House], quotes from the "Epilogue"

papers and books published that described and discussed *real* biological sex differences. A few of these works were:

Males and Females by Corinne Hutt, 1972;

Psychology of Women by Judith M. Bardwick, 1971;

The Development of Sex Differences by Eleanor E. Maccoby, 1974;

The Psychology of Sex Differences by Eleanor E. Maccoby and Carol N. Jacklin, 1974;

Gender Differences: Their Ontogeny and Significance, edited by C. Ounsted and D.C. Taylor, 1972;

Man & Woman, Boy & Girl by John Money and Anke A. Ehrhardt, 1972;

"Innate Masculine-Feminine Differences," W.J. Gadpaille, *Medical Aspects of Human Sexuality, Feb 1973*

Differential Psychology by Anne Anastasi, (3rd ed.) MacMillan: New York, 1965.

Sex Differences in Mental and Behavioral Traits. *Genetic Psychology Monographs*, Josef E. Garai, and Amram Scheinfeld. 1968, Vol. 77, 2nd half, pp. 169-299.

The Psychology of Human Differences (3rd. ed.). Leona E. Tyler, Appleton-Century-Crofts: New York, 1965.

Research into the Physiology of Maleness and Femaleness. *Archives of General Psychiatry*, Warren J. Gadpaille, 1972, Vol. 26, pp. 193-206.

The realists (man and women) recognize the reality of biology and work within it in order to form a fairer environment. We will examine the real reality of sex-differentiated biology and how much it does dictate many aspects (not all) of our sex roles.

From the above comprehensive works, other specialized papers, and more recent works[82] we will present to you important sex differences in the biology and the behavior of human beings.

In order to understand the mistakes of radical feminism we must understand real sex/gender differences and the evidence found that indicates strongly that they are mainly biologically caused and not

---

[82] see partial list in our "Brain Organizational Differences & Hormones" in this chapter

culturally caused. We can understand why some feminists *need* these gender differences to be caused by socialization, for then, it may be possible to change them. But if these differences are biologically caused it would be more difficult, if not impossible to change. Of course, even if they are biologically caused, a society could/should attempt to culturally change *attitudes* that are unjust against women, and change unfair attitudes against men that some hold; incorrect attitudes are not the sole domain of men.

The apparent low status of women's work or her general status *per se* is one unfair aspect that must be changed, not only because it is unfair, it is inherently *wrong*. Men and women should be co-equal partners in marriage and society with each playing their complementary roles and each receiving equal status. The unfair attitude toward women will be examined in a later chapter.

What is Sex?

What is sex? The word "sex" in English comes from a Latin word *secare* which means to cut or divide. And this usage probably originated from the biblical rendition of the genesis of mankind.

> Genesis 2:20 And Adam gave names to every animal, and to the fowl of the air, and to every beast of the field; **but for Adam there was not found a helpmate corresponding**[83] **to him.**
>
> 2:21 And the Becoming-One God caused a deep sleep to fall upon Adam, and he slept: and **one [fem gender] he took from his side**, and closed up her place with flesh; [so she was a part of him to begin with]
>
> 22 And the side [fem] which the Becoming-One God had taken from man, made he a woman, and brought her unto the man.
>
> 23 And **Adam said, This is now bone from my bones, and flesh from my flesh**: she shall be called Woman, because she was taken out of Man.
>
> 24 Therefore shall a man leave his father and his mother, and be joined unto his wife: and **they became one flesh**.
>
> Get 5:2 **Male and female created he *them***; and blessed *them*, and called *their* name Adam, in the day when they were created.

---

[83] The Hebrew word means "in front" "opposite" or "corresponding" or even "complementary" so translating "helpmate corresponding" can mean *complementary partner* or *corresponding partner* or *opposite partner* when taken in context

Differences. Notice according to the Bible the two (male & female) are separate/opposite helpmates: two in one. They help each other. They are partners. They being two, have one name (Adam — mankind) and they together are one flesh. As we know **two can become one** through sexual intercourse. The resulting newborn (one from two) has one half of each of their parents' chromosomes. In this account the first woman was cut from the side of man. She was taken from him or divided from him. Science, as we will see, also manifests to us that there is a division, that there are differences (opposite/complementary/corresponding) between males and females. The word "sex" indicates a division of mankind or a separation of mankind. There are elements or characteristics that separate females from males. "Sex" has to do with these differences. Sex is difference, not likeness. Although males and females are alike in some ways (both have two arms, two legs, etc.), both are also unalike in some ways (genitals/hormones levels/brain wiring), and in totality both are unalike in each and every cell (XY [males] or XX [females] chromosomes). The things in which they differ are the divisions between them.

Sex is Dyadic/Complementary. Often the male and female differences complement each other. Although there is a sexual division in mankind, each complement the other and both *together* make up a complete functional unit. Mankind can best be viewed as *dyadic*: two incomplete components (male and female) that complementary make up a complete unit. The sexes are incomplete apart; they are complete together. Mankind continues only because the two components unite into a union of complementary roles and responsibilities in which, through physical and mental intercourse, they produce and sustain their kind. Marriage is the normal method in which the sexes join together. And thus the biblical verse, "For this reason a man will leave his father and mother and be united to his wife, and the two will become one flesh." (Mark 10:7-8)

Males and females are relatively and/or absolutely different biologically:

>in sexual organs,
>in sexual functions,
>in brain organization and activity,
>in form,
>in dimensions,
>in muscular strength,

in energy production and energy expenditure,
in heart pulse rate,
in respiration rate,
in cell composition,
in maturation rate,
in bio-chemical ratio,
in muscle-fat ratio,
in chemical (hormonal) ratio variability,
in homeostasis,
in weight variability,
in red and white blood cell ratio,
in internal influences on behavior,
in intrinsic tendencies of behavior,
in perceptual distractibility,
in erotic perceptual arousability,
in morbidity,
in mortality,
and in other factors.

Totally Different. In fact, males and females are in one sense entirely different: *each* cell of the normal male has the XY sex chromosomes, while *each* cell of the normal female has the XX sex chromosomes. Not only this, but also each sex's differentiate cell reacts somewhat differently to hormones and medication. This last point is one huge mistake pharmaceutical companies made in the past in testing drugs: they tested mostly males instead of an equal number of females because they didn't understand that drugs can react in sex differentiated ways.

Sex Differences From Conception Onward

Sex differentiation begins early in the development of the sexes — from conception. In each of our cells we have 46 chromosomes, 23 being contributed by each of our parents. The chromosomes align themselves in 23 pairs, one from each parent. Twenty-two pairs are called *autosomes.* They carry genes that determine the individual physical features we have. One pair of chromosomes are the sex chromosomes and they determine which sex we are. If we have the XX chromosomes we are female. If we have the XY chromosomes we are male. The sperm or egg of the male and female each has only one-half of the normal chromosome count, or 23 chromosomes. Each of the female's eggs contains one X chromosome. Half of the male's sperm contain the Y chromosome and the other half the X chromosome. Thus, the mother always provides the X chromosome; the father provides either an X or a Y chromosome. Therefore it is the father's sperm type that fertilizes the X-bearing egg and determines the sex of the child. If the father's sperm contains a Y chromosome the child will normally develop into a male with XY chromosomes in each of his cells. He is an XY genotype. If the father's sperm contains an X chromosome the child will normally develop into a female with XX chromosomes in each of the female's cells. She is an XX genotype. There are *rare* cases of XO (female: Turner's syndrome), XXX (female: in many cases may be mentally retarded), XYY (male: overly impulsive, aggressive), and XXY (male: Klinefelter's syndrome — breast enlargement, small penis and testes) genotypes.

Critical Period — 7th/8th Week. Up to and through the sixth/seventh week of the child's development in the womb the fetus is not additionally sex differentiated beyond the difference in the sex chromosomes. But after the sixth/seventh week of development it then begins to sexually differentiate into a male or female. This differentiating is determined by the relative presence or relative absence of the male hormone in the developing fetus. To understand this more, we need to next look at the influence of hormones. Some newer studies[84] say it is the eighth week that is the critical period.

Sex and Hormones

The word "hormone" comes from a Greek word meaning, to stimulate, or excite. Hormones are secreted from the endocrine glands. The sex endocrine glands are called the gonads. In the male, his gonads are his testes; in the female, her gonads are her ovaries. Also the adrenal and the pituitary glands are glands that are sexually differentiated in certain ways.

The male sex hormones are collectively called androgens. These have a masculinizing action. The main male hormone of the androgens is testosterone, which is mostly secreted from the cells of the testis. The testis, not only produces male hormones, it also produces sperm that enables a male to become a father through impregnation of a female with the sperm.

---

[84] Arnold, AP. "Sex chromosomes and brain gender, *Nat Rev Neuosci*. 2004; 5: 701-8.

The secretion of androgens in the male begins prenatally and is relatively decreased postnatally, but increases in level throughout later childhood until puberty, when there is a constant high secretion of male hormones, approximately 10 times the female's level. The level of androgens begins to drop after 50 years of age.

In the female, the ovary secretes two distinctive sex hormones — the estrogen and the progesterone. The principal estrogen produced is estradiol that is made from the cells of the ovary. The ovaries also produce the ova (egg) that makes it possible for females to become mothers. The estrogens help in the growth and maturation of the female reproductive parts — the vagina, uterus, and oviduct. These hormones help in the development of the mammary glands. Progesterone is essential for various aspects of pregnancy. Females' estrogen and progesterone high levels begin with puberty, vary somewhat during their menstruation period, vary greatly during pregnancy, and decrease dramatically after menopause to about one ninth the level of a menstruating female. The level of estrogen and progesterone of females vary over their menstrual period. When she is pregnant the levels significantly increase until at the end-of-pregnancy estrogens are 10 times above normal, and progesterone is about 100 times above normal. Contrariwise the level of hormones of males is relatively steady.

After menopause women seem to become more aggressive and assertive, their voices deepen and they grow some facial hair because their adrenal glands are still producing small amounts of androgens, but their estrogen level has dramatically decreased. The ratio of their hormone level changes; estrogen decreases its influences, and they thus change.

Male & Female Hormones

Often androgens are called the male hormones and estrogens are called the female hormones. This isn't technically correct since females also produce testosterone (an androgen), but at a much lower level than males. Males also produce some estrogens, but at a much lower level than females. Therefore, androgens are the male hormones because males have a higher level of them than females. Estrogens are the female hormones because females have a higher level of them than males.

Hormonal Interaction and Control

The sex hormones do not act alone. They interact or interplay with the whole body in various ways. However, they are more under the influence and control of the pituitary gland and the hypothalamus of the brain than other bodily parts. What makes this pertinent is that the hypothalamus is sexually differentiated.

Brain Organizational Differences & Hormones

One of the most important discoveries concerning sex differences, had to do with innate differences between the male and female brains.[4] John Money's 1972 book, *Man & Woman, Boy & Girl*, gave a well-documented explanation of the fact that there is a difference between the male and the female brain. But this book gave too much attention to the neutrality-at-birth theory of gender identity (see Chapter 3) that only postponed the truth of brain sex for the general public. After Money's undeserved influence from his misguided neutrality-at-birth theory in the 1970s and 1980s (even though he documented real sexual differences in brain organization and function), there came scientific articles and books on biological sex differences in the brain. One somewhat popular book, *Brain Sex*, and its three-part TV series also called *Brain Sex* was published and aired in 1991-92. Both of these documented the evidence for the innate brain differences between the sexes.

Our quick review that follows uses Money's book, other newer research papers (Breedlove, 1992, "Sexual Differentiation of the Brain and Behavior." In *Behavioral Endocrinology*), and such popular reviews of pertinent literature as the 1983 book *Sex and the Brain*, and the 1991 book, *Brain Sex* [5], and the two books by Louann Brizendine called *The Female Brain (2006)* and *The Male Brain* (2010) as well as the interesting book by Katty Kay & Claire Shipman, *The Confidence Code* (2014), on women's apparent lack of confidence as compared to men. Another book (2008) on sex differences in the brain, *Sex Differences in the Brain: From Genes to Behavior*, by Elizabeth Young, *et al.* is more of a survey of literature, not very readable by the average reader, with details on chemical-hormonal differences, animal sexual differences, disease differences and with reports of some human studies. Also a 2010 survey of gender differences in the human brain can be found on the Internet on the benthamopen.com website called, "Gender Differences in Human Brain: A Review," by Zeenat F. Zaidi with 276 footnotes.[85]

The Differences are Manifold

"Males and females differ in such traits as their averages, extremes, permanence, temporal qualities, susceptibility to disease, and in their functional impact."[86]

The hypothalamus is a key brain structure. The differentiated brains of the sexes have something to do with the brain's hypothalamus. The hypothalamus is a key subcortical structure at the base and center of the brain. It is a *key* brain structure because it has an important influence on emotional behavior, thirst, hunger, temperature regulation, metabolism regulation, and even motivation. [6] It follows then that any sex difference of the hypothalamic action is important to our understanding of the difference between each sex. As Corinne Hutt put it in her book, *Males and Females*:

> Because the hypothalamus is such an important control center and because many parts of the subcortical brain are intimately connected with each other, it is unlikely that other non-sexual functions controlled by the hypothalamus remain entirely unaffected by its differentiation according to sex.[7]

From Brizendine's, *The Female Brain*, we read:

> In both males and females, testosterone is the chemical fuel that gets the brains's sexual engine going.... Testosterone revs the hypothalamus, igniting erotic feelings and arousing sexual fantasies and physical sensations in the erogenous zones. The process works the same way in men and women, but there's a huge sex difference in the amount of testosterone that's available to "turn on" the brain. Men have on average ten to one hundred times more testosterone than women.[87]

---

[85] http://benthamopen.com/journal/render-fulltext.php?articleID=TOANATJ-2-37#R162; or go to my web site for a PDF copy (parisburg.com/authors/DolenDocs/GenderDifferences.pdf).

[86] Young, Elizabeth; Becker, Jill B.; Berkley, Karen J.; Geary, Nori; Hampson, Elizabeth; Herman, James P. (2007-10-31). S*ex Differences in the Brain:From Genes to Behavior* (Page 15). Oxford University Press. Kindle Edition.

[87] Brizendine, See 28a, p. 89

The hypothalamus is differentiated prenatally. By the action of androgens from the prenatal infant, at a critical stage in fetal development (7th/8th week), the hypothalamus of the infant is organized into a male pattern. Without the introduction of androgens at this critical stage, the fetus develops a female hypothalamic organizational pattern. The main known sex difference of the hypothalamus is that the male's hypothalamus works in a noncyclic manner, while the female's hypothalamus works in a cyclic manner. [8] The cyclic pattern of the females has to do with the release of neurohumoral substances from the nerve cells of the hypothalamus. These neurohumoral substances in turn regulate the nearby pituitary gland.[9] And the pituitary gland in turn regulates many functions of the other endocrine glands of the body like the testes and ovaries. Thus, because the female hypothalamus functions cyclically, because it affects her ovaries in a cyclic manner, and because her ovaries produce hormones, it follows that her hormonal levels vary in a cyclic pattern, which they do.[10] This fact, directly and indirectly, helps to create cyclic moods or temperamental changes in women.[11] Since hormones are chemical stimulators this latter statement makes sense. Each hormone affects the body in certain ways. Most everyone notices what the caffeine in their coffee does for them in the morning. For radical feminists to deny the stimulating influences of the hormones on the body, is like them denying the affects of caffeine or any other chemical stimulator. The newest research in neuroendocrinology tells us the hypothalamus of males and females acts differently. And because this area of the brain is a control center for other parts, this is important knowledge for understanding sex differences in behavior.

Male or Female Brains from the Womb. It has been shown that mating behavior is dependent upon whether the participants have either a male or female "brain."[12] And such non-mating behavior as physical aggression, maternal behavior, energy expenditure, choice of clothing, choice of career, erotic perception, romantic fantasies or dreams, and so on are also related by whether the person has a male or female brain.[13] Male and female brains develop differently in the womb and become "wired" differently prenatally. There is a short critical period in the seventh/eighth week of development of the fetus. If male hormones are introduced into the fetus at this critical period, the child develops into a male with a male brain and genitals; otherwise it develops into a female with a female brain and genitals.

4. Sex Differences and their Implications

There are males, as well as females, because of androgenic activity in the prenatal infant. [14] Without this hormonal activity nature would only produce females. The undifferentiated fetus, whether the fetus is of XX or XY genotype, will always change into a female unless androgens are introduced into the prenatal infant by its fetal gonadal hormones, or exogenous sources, at the critical period in its prenatal growth. Although no ovarian hormones are necessary in order for nature to produce a female, there is evidence that they do influence how the female brain is developed.[88] In order for nature to produce a male, testicular secretions must be present and acting in the fetus at the proper level and proper time. Therefore, androgens are a very important ingredient in the formulation of the sexes. Without its influence in the prenatal infant, there would only be females.

Normally the XY chromosome fetus (male) automatically develops testes in the seventh/eighth week of development, and they in turn produce the male hormones that immediately begin to develop a male brain and other male features. When a child is born they already have a male or female brain. Their brains are "wired" male or female beginning in the seventh week of development in the womb. There is proof that maternal behavior is wired into the brain prenatally. (see below) There is proof that the sexes use the left and right hemispheres of their brains differently and have different physiological connections between them. [15]

From birth and onward males and females use the left and right hemispheres of their brains differently, different levels of hormones affect them differently, **and thus the sexes throughout their lives have different**:

>   spatial ability (manipulating shapes of things in the mind)
>   verbal ability
>   hearing ability
>   visual ability
>   reaction to pain
>   tasting ability
>   smelling ability
>   memory ability

---

[88] Fitch & Denenberg, "A Role for Ovarian Hormones in Sexual Differentiation of the Brain," 1997: Cambridge U Press [pre-press text]

emotions

anger

aggression [16]

The biological differences help, directly and indirectly, to make males and females behave and relate differently. They perceive their environments differently, in a sex-differentiated way.

Because their brains are "wired" differently, they see the world differently and learn differently. It is the interaction or interplay of the external environmental influences with the persistent internal biological influences that produce yet more differences in male and female behavior. Yet, the main *cause* of sex behavior is biology. Knowing this we can see why the sexes don't seem to understand each other:

> Why do men do that? Why do women do that? Men are so stupid. Women are so dumb.

What is really being said here is that "the other sex does not think and react the same way I do," so they must be "stupid" or from a different planet. No, their brains are "wired" differently and hormones cause them to be more inclined to think and act in different ways to the *same* simuli. Each sex needs to be cognizant of this and react and think about each other accordingly.

Intellectual Ability Differences — IQ Tests

Because of the brain differences there are sex differences in mental ability. IQ tests, which are supposed to measure intellectual ability, are constructed to eliminate sex differences in total scores:

> The tests now in use were deliberately designed and arranged to suppress sex differences and to make the average I.Q.'s of boys and girls come out equal.
>
> From the time the tests began [1916, now known as the "Stanford-Binet test"]... it was found that there were some questions and problems on which girls did consistently better, some on which boys did better. The assumption was that these differences were due almost entirely to conditioning — an assumption which still prevails.
>
> ....They threw out all items in which those of either sex proved to be markedly superior to those of the other sex. Yet, try as they might, in the remaining questions and problems, shades of sex difference in superiority persisted. So,

continuing, the tests were arranged to balance the "girl plus" items with "boy plus" items in such a way that the total average scores would come out even.

On separate items, the average scores for each sex are often radically apart. Moreover, even where the same mark is secured on a specific item, it is often found that the minds of boys and girls have worked differently toward reaching the solution, just as two persons might get to the same destination by taking different roads.[89]

Therefore, individual test items that show *large* sex differences in response are excluded entirely. This procedure is based "on the assumption that sex differences on such items may be specific to the task in question and may simply reflect differences in experience and training. Among the remaining items, those slightly favoring girls were balanced against others which favored boys to an equal degree." [17] When an IQ test shows generally equal total test scores for each sex, it merely shows how well the test was constructed to rule out sex differences:

In the 1937 Stanford Revision, a Definite, and probably successful effort, was made to eliminate tests for which either sex showed a preference; the authors of the revision did not consider it 'fair' for one sex to do better than the other. Thus all study of sex differences by means of the Stanford revisions is infeasible.... [18]

Sex/Gender Differences in Verbal Ability, Number Ability, Spacial Ability Found in IQ Tests

Even though IQ tests are purposely constructed to eliminate sex differences, sex differences do appear in factor analysis, that is analysis of the Word Fluency factor and Space factor within the test. Through factor analysis of the IQ tests, various sex differences in the kinds of intellectual ability have been manifested:

(1)Verbal Ability or Word Fluency. Females from infancy to adulthood show greater ability in verbal functions like word fluency, spelling, grammar, speed of reading, etc.[19] But in verbal reasoning and vocabulary, both sexes have about the same ability.[20] For example, in studies of eighth and ninth-grade boys and girls in 1944-46, girls outscored boys in the Word Fluency part of the tests

---

[89] *Women and* Men, Amram Scheinfeld [151], pp 84-85; see also Bibl # 61, pp. 173. 196ff

75 to 68, 61 to 55, and 55 to 51 while all other factors except Spatial were within a point or two. [21]

Verbal Skills of Women. While boys focus on *objects* (visual stimuli), generally girls of the same age focus their attention on each other in interpersonal activity, or on conversation about other persons. [22] Females are better at verbal activity. Females have greater verbal fluency skills from infancy on.[23] Being verbally fluent is being able to easily put into words what one has to say or write. Girls learn to speak earlier than males. They do better in school in literature, essay writing, grammar, spelling, and foreign languages. Girls have far fewer speech problems than males at all ages. Males generally do reach, and some slightly surpass, females' level of ability in vocabulary, verbal comprehension, and verbal reasoning in their late teens, but do not become as good as the average female in other verbal areas unless they put much effort into it. Generally, females are the master of language mechanics and verbal fluency.

This greater verbal ability of females shows itself in their greater interest in social or interpersonal aspects of life. Females are more interested in people than in objects. They have greater affiliational needs than males, and are motivated to do things more because of social reasons (the praise, acceptance, or recognition it will bring from others of importance to her) than because of the thing itself. [24]

Why are women more verbal and social than men? It is in the chemical rewards they get from talking and being intimate and social:

> Connecting through talking activates the pleasure centers in a girl's brain. Sharing secrets that have romantic and sexual implications activates those centers even more. We're not talking about a small amount of pleasure. This is huge. It's a major dopamine and oxytocin rush, which is the biggest, fattest neurological reward you can get outside of an orgasm. Dopamine is a neurochemical that stimulates the motivation and pleasure circuits in the brain. Estrogen at puberty increases dopamine and oxytocin production in girls. Oxytocin is a neurohormone that triggers and is triggered by intimacy. When estrogen is on the rise, a teen girl's brain is pushed to make even more oxytocin — and to get even more reinforcement for social bonding.[90]

---

[90] Brizendine, 28a, p. 37

4. SEX DIFFERENCES AND THEIR IMPLICATIONS

An Example of Bias. Maccoby and Jacklin, in their book *The Psychology of Sex Differences,* conclude that it is an unfounded belief that girls are more social than boys.[25] They state that both sexes are equally interested in social stimuli as compared with nonsocial. Maccoby and Jacklin mention studies where infants were tested on how much attention they directed to social or nonsocial stimuli. Little or no sex difference in attention was found in these studies. But what did they consider social stimuli? The answer is that various representations of human faces such as simple line drawing of faces, or clay facial masks, or photographs of faces, or actual real faces were considered social stimuli.[26] Nonsocial stimuli included such items as checkerboards, or geometric forms, or random shapes. It seems incredible that Maccoby and Jacklin could use such a dubious interpretation of studies to try and disprove the conclusions of numerous authorities as to the fact that females are more socially oriented than males. The subjects of the experiments mentioned by Maccoby and Jacklin were infants, most of which were under six months of age.

How social are such infants? When they cannot talk, how can they be sociable? Are the faces presented to these infants, social stimuli or visual or spatial stimuli? It seems Maccoby and Jacklin's biased beliefs [27] may be interfering with their judgment. Even in their book, *The Psychology of Sex Differences*, they admit that: "The summaries of earlier research presented in *The Development of Sex Differences* (1966) indicated that women and girls showed more interest than boys in social activities and that their tastes in books and TV programs were more oriented toward the gentler aspects of interpersonal relations...."(p. 214) Although Maccoby and Jacklin's book is important in many respects, one should be careful about some of their conclusions. They have a tendency to be biased in a radical feminist direction.

(2)Number Ability. In the lower grades there isn't much difference, but when computational arithmetic problems change into reasoning arithmetic problems of higher mathematics, then males excel.[28]

(3)Spatial Ability. Males have shown consistent superiority in the manipulation and judgment of spatial relationships.[29] Spatial ability manifests itself in mechanical ability, skills involving space relationships (architects), practical ability, and higher mathematics. [30] For example, in studies of eighth and ninth-grade boys and girls in 1944-46, boys outscored girls in Spatial Ability 83 to 69, 42 to 33,

and 40 to 31 while all other factors except Word Fluency were within a point or two.[31]

Males' spatial ability manifests itself in their better performance in the solving of mazes and puzzles, in areas of mechanical ability, in higher mathematics, and so forth.[32] Males have better visual acuity in all ages studied as shown from a study of 17,500 California drivers whose ages ranged from 16 to 92.[33] Males as early as 14 weeks were found to learn much better to fix their attention on objects because of visual stimulus reward versus an auditory stimulus reward.[34] All of these aspects plus the visual erotic difference indicate that men are influenced and interested more in visual stimuli than women. This is also manifested in pre-school children where "when boys are gathered together it is generally as a group of three or more, and their attention tends to be focused on some activity or on objects...."[35]

Parenthetically, the study of individuals with Turner's syndrome seems to indicate an androgenic foundation for the greater spatial ability of males. In Turner's syndrome, no hormones in the womb reach and affect the brain to sexually masculinize it because such individuals have no functional gonads. Persons with such a condition sexually differentiate as females who are much more feminine-behaving than the average girl. It is the fact that such individuals are greatly retarded in their spatial ability [36], even though they are normal or slightly above in their verbal ability, that seems to indicate a prenatal-androgenic foundation of spatial ability. Spatial ability sex differences appear at puberty, fluctuate during the menstrual cycle, and appear to correlate with changes in estrogen levels.[91]

To Conclude. The verbal, number, and spatial abilities, are the main and more provable mental ability differences between the sexes.[37] Whether any of these abilities are superior to others is merely opinion. We call someone intelligent because he or she has certain mental abilities or behavioral abilities that we deem important. Some people include practical ability with academic ability when evaluating intelligence, and some do not. But the fact remains, IQ tests do not tell us which sex is smartest because the tests were designed to be sex neutral.

---

[91] Halpern, *Sex Differences in Cognitive Abilities*, London: 1992; Hampson, Brain and Cognition 14:26-43, 1990; Hampson & Kimura, *Behavioral Neuroscience* 102: 456-459, 1988

Statements such as, "spatial ability is only weakly correlated with measures of general intelligence,"[38] by the psychologist Maccoby are revealing their bias. Why? Because general intelligence is measured by IQ tests. If IQ tests show spatial ability is weakly correlated to general intelligence, it is because these tests are constructed that way, not because spatial ability has nothing to do with intelligence. Most IQ tests, in fact, have few questions that deal with difficult spatial problems. If there were more questions on spatial ability, then spatial ability would be highly correlated to general intelligence tests. It all depends on how the tests are constructed. Even though IQ tests are made to be as sex neutral as possible, they still manifest sex differences in verbal, math, and spatial ability, which is projecting the innate sex difference of the brain in each sex.

Other Perceptual Differences

Breaking Set, Field-Dependence & Independence. Another aspect of the brain differences between the sexes are the perceptual differences in field-dependence, field-independence, and breaking-set. Females are more field-dependent.[39] The female has a more *global* approach to the visual perception of spatial relationships than the male; she perceives the visual stimulus and its surrounding field as a whole interconnected reality.[39a] She sees the whole room. The male is more field-independent.[39] He concentrates on the visual stimulus itself, and is less affected by the surrounding visual field. He sees and focuses on a part or parts of the room. Males tend "to deal with the field in an active, analytical fashion."[40] They manipulate the parts of the visual field in their head to a greater degree than females. Females tend "toward passive acceptance of the field."[40] Males are more willing and able to break away from the total visual picture presented, and to actively move the elements of the picture in their minds. This difference of perception shows itself as early as three or four years of age.[41]

This field-independence ability of males manifests itself also through males' greater "set-breaking" ability as noted by Garai and Scheinfeld:

> "Men and women appear to be equally susceptible to the adoption of a certain fixed attitude or set toward the solution of a problem, but men are able to abandon a once adopted set more easily than women, when a new and different approach is required for the successful solution.[42] This male

superiority in set-breaking capacity or the restructuring of the perceptual field was also observed by Sweeney[43] and Kostick.[44] The latter found school boys more able than school girls to 'transfer' their knowledge and to apply their skills to new situations in science, a 'masculine,' and home economics, a 'feminine' area of interest."[45]

This set-breaking ability is one reason why males generally do better at solving a problem where breaking set or thinking in a new way solves the problem.[46]

Perceptual sex differences manifest themselves in the sexes' different fields of interest. For example, the greater mathematical ability of males is more than likely because math is a nonverbal language. Since men do not have as high a social need as women, they can interest themselves in math, while women, because of their social need, will be more interested in studies where they can use their knowledge in interpersonal verbal contact. Thus, since higher math can only be communicated visually (backboards, textbooks, computers), women are less inclined to study it with the great interest needed to excel in higher mathematics. Women are more inclined to study *verbal* languages, or *social* science, or history of *people*, or economics of the home or *family* instead of visual languages (math, architectural drawing), or physical science, or history of ideas, or economic theory. When women enter male-dominated fields like physical science or economic theory, they often specialize in the more feminine aspects of them, like teaching, instead of research and theory development.[47]

The results of this sex-differentiation is that a social framework is set up where young females and young males identify with traditional sex occupational divisions and unconsciously set their goals to agree with these divisions. Thus, more males choose "male" occupations, and more females choose "female" occupations than would occur if there were no subtle social pressures to do so. But this is actually a good thing because if young females identified and studied for male occupations, they would less likely be fulfilled (be happy) with their occupation than if they chose a female occupation. A female occupation is one where their biological tendencies are in part or in full satisfied.

Verbal and Spatial Ability in Higher Mathematics, Engineers, Scientists and Spatial Ability

To quote Camilla Persson Benbow concerning her conclusions from her eight year study with Julian Stanley of Johns Hopkins

4. Sex Differences and their Implications

University, "Sex Differences in Mathematical Ability: Fact or Artifact?" published in the journal *Science* in 1980:

> We're just beginning to have evidence that when there are two equally valid approaches to a problem, via words or via images, females tend to choose the approach through words and males the approach through images. Now the approach through images—which are visual-spatial and right-hemisphere[92]—just happens to be much more effective, especially in higher mathematics, than the approach through words. Look at the way mathematicians are forced to talk to one another, through symbols on a blackboard. And so I think that their right-hemisphere approach naturally favors males. From the beginning they're less verbally oriented then females—more oriented to things, to objects in space. They're less dependent on context in their visual-spatial skills—this can be seen cross-culturally from the age of four onwards. And they're more abstract.
>
> This may help explain, too, I think, why men are over-represented in certain disciplines in science, something we've also been studying. To be a good physicist or engineer, for example, requires not only mathematical reasoning ability but also skill in three-dimensional visual imagery. And that's probably why few women are found in these fields. To be a good scientist, at all, in fact, seems to require a set of qualities more characteristic of men than of women—spatial ability, a low social interest and an absorption in things. Let's face it, human males like to manipulate *things*—from Tinkertoys to the cosmos.... Females are more communicative, more sensitive to context and more interested in people. And perhaps that's why there are so many more women in fields like biology and psychology: like me.[93]

---

[92] New fMRI studies seem to question the popular ideas of left-right brain, and gender differences. Nielsen, Zielinski, Ferguson, Lainhart, Anderson, "An Evaluation of the Left-Brain vs. Right-Brain Hypothesis ..." Aug 14, 2013, Plos One; and see Ingalhalikar, et al. "Sex differences in the structural connectome of the human brain," Sept 9, 2013, PNAS

[93] Quote from *Sex and the Brain*, Chapter 5, pp. 68-69, Warner Books Edition, July 1984. "Sex Differences in Mathematical Ability: Fact or Artifact?," *Science*, December 12, 1980.

Erotic Perceptual Differences

Because of the differences in the male and female brains, and the different levels of hormones, there is a distinctive sex difference in visual erotic perception. In such magazines from the 1970s as *Playgirl*, nude men were presented for the liberated NEW women. According to some, women have been conditioned by their puritanical parents and society away from enjoying the opposite sex's nudity, as men enjoy their opposite sex's nudity. But is this the real answer?

Hormones and Arousal. One reason human males get excited by female nudity is because of hormonal differences. Men have a higher ratio of androgens to estrogens, while women have a higher ratio of estrogens to androgens. Androgens have been found to be an erotic arousal hormone in males and females, while estrogens are an erotic tranquilizer. Androgen is a libido-enhancing hormone for both sexes. [48] By contrast, estrogen is a functional castrating agent in the male. It thus may lower the intensity of libido and lead to a decrease of sexual initiative and activity. In effect, it is an erotic tranquilizer in the male.[49]

Thus, one reason boys get excited over the female form after puberty, and one reason sex is often on the minds of boys,[94] is because of the increase in their androgenic level at the time of puberty. The hormonal changes of puberty make males more sensitive to erotic stimuli than females. The lower level of androgens that females do produce in their bodies also erotically induces sexual excitement in them, but at a lower level than men.

Visual Arousal. Although "it is possible for a woman as well as a man to respond to the visual stimulus of a lover or potential partner in sex,[50] her imagery of arousal will however, tend to be different...."[51] Her sexual arousal is built around romantic ideas of *him* "reacting to her and wanting her — of wanting to hold, caress, and kiss her."[52] In women, the memory of past sexual experience, in real or book form, is important for her arousal by a nude male.[53]

---

[94] "85 percent of twenty to thirty year-old males think about sex many times each day and women think about it once a day — or up to three or four times on their most fertile days." — 28a, p 91. Studies of college men indicate they think of sex 19 times a day on average, while college women may think of it 9 times a day: Fisher, Moore & Pittenger, "Sex on the Brain: An Examination of Frequency of Sexual Cognitions as a Function of Gender...," Vol. 49, Issue 1, 2012, *The Journal of Sex Research*. Other studies indicate women's sexual thoughts are tied to their cycle.

4. Sex Differences and their Implications

The more a woman has positive sex experience, the easier she is sexually turned on by a male. Women are aroused not by men's bodies *per se*, but by her imagery of *his* desire for her. Women look at a desirable male and think of how nice it would be for him to make love to her, not how nice it would be to touch him. (Although she may learn to think in the latter way through positive sex experience: she learns that he likes to be touched in certain ways; and thus she learns to like these activities *because* he likes them, and she wishes to please him.)

Women as Sex Objects. Conversely, a woman's body can turn on men sexually in and of itself. When looking at a desirable woman, a man thinks of how nice it would be for him to caress, handle, and fondle her. He objectifies her. He makes her the object of his attentions. He uses his better spatial ability, his field-independence ability to objectify the female form in his mind. But women do not objectify men in this way. Women see themselves as objects of male's desire. This is the reason they like, "I love you," sentimental gifts (as outward signs of love and desire for them), and passion for them.

This erotic sex difference in perception is one reason women care for their appearance. This sex difference is also why such women as the feminist Germaine Greer think men with clothes on are sexier than nude men: "My favorite Renaissance sex object is fully, even heavily dressed."[54]

Hierarchy of Desirability For Both Males and Females. Women like men who are desirable in a social way, more so than men who have well-built bodies, although a man's height is important to many women who can't see themselves with a shorter man. Women look at the way a man moves, or handles himself in various social contexts more than how his face and body are formed. Does he have an intelligent mind and social status? Good. Does he dress or act in socially desirable ways? Good. And if he is handsome, that's added desirability, not primary desirability.

On the other hand, men look *initially* at a woman's appearance more so than at her social or mental attributes. This doesn't mean he doesn't weigh her social and mental attributes. It does mean males are cognizant first of a woman's physical appearance, and second of her behavioral manifestations, especially younger men. More men as they age appreciate women's non-physical beauty.

A short synopsis of a rather comprehensive review on women's and men's preferences and choices of mates and marriage partners by <u>David C. Geary</u>, Jacob Vigil, and Jennifer Byrd-Craven, "Evolution of Human Mate Choice" is given here:

> In particular, we focus on women's and men's preferences and choices of mates and marriage partners, and invite the reader to judge for himself or herself the utility of this approach.... [p. 27]
>
> In classical literature and in romance novels, the male protagonist is almost always socially dominant, wealthy, and handsome (Whissell, 1996). Indeed, a preference for an attractive mate makes biological sense (Fink & Penton-Voak, 2002; Gangestad, 1993; Gangestad & Buss, 1993). Not only are handsome husbands more likely to sire children who are attractive and thus sought out as mating and marriage partners in adulthood, but these men and their children also appear to be physically healthier than other men and their children (Gangestad, Thornhill, & Yeo, 1994; Grammer & Thornhill, 1994; Singh, 1995a; Thornhill & Gangestad, 1993, 1994). In other words, the physical attributes that women find attractive in men are indicators of the man's physical and genetic health (Gangestad & Simpson, 2000).... [p. 32]
>
> In short, most women prefer monogamous marriages to wealthy, socially dominant, and physically attractive men, and want these men to be devoted to them and their children.... [p. 39]
>
> The primary differences are that men are more focused on the physical traits of a long-term mate and less concerned about her cultural success or her potential for cultural success (Buss, 1989; Li et al., 2000). In theory, men should have evolved to focus on those physical attributes of women that are predictive of their reproductive potential, specifically their ability to conceive, carry, and birth healthy children. These traits include age, body mass index, waist-to-hip ratio, and breast symmetry, among others (Andersen et al., 2000; Singh, 1993a; Møller et al., 1995; Zaadstra et al., 1993). As predicted, men do indeed focus on these traits when judging the attractiveness of women as potential short-term and long-term mates (Kenrick & Keefe, 1992). [p. 39][95]

---

[95] Review found in *The Journal of Sex Research*, Vol 41, No. 1, Feb 2004, pp. 27-42; <u>http://web.missouri.edu/~gearyd/MatechoicePDF.pdf</u>

4. Sex Differences and their Implications

To understand the complexities of male and female choices for the opposite sex, one should read the above study and read more popular books such as *The Male Brain* (2010) and *The Female Brain* (2007) by Dr. Louann Brizendine or *The Essential Difference: Male and Female Brains…* (2003) by Simon Baron-Cohen.

How does an erotic female sexually arouse women? Many men mistakenly think that women are aroused by what they are turned on to sexually. To men, a nude female is sexy. Thus, some think a nude female is also sexy to women. And sometimes it is for women, but in a different way. Some think that women who are "turned on" by nude women, are manifesting their lesbian tendencies.[55] But this isn't the case. John Money and Anke Ehrhardt explain this:

> When he reacts to a sexy pinup picture of a female, a man sees the figure as a sexual object. In imagery, he takes her out of the picture and has a sexual relationship.
>
> The very same picture may be sexually appealing to a woman, but that would not mean she is a lesbian. Far from it. She is not in imagery bringing the figure toward herself as a sexual object, as a man does. She is projecting herself into the picture and identifying herself with the female to whom men respond. She herself becomes the sexual object.[56]

Therefore, the way a woman can be sexually aroused by an erotic female is to *project* herself as the girl in the picture, film, or live setting. And the way men are sexually aroused by an erotic female is to *objectify* the girl — he makes love to her in his mind as he looks at the woman in a picture, film, or in a live situation. In this section we are referring to women who have *not* been androgenized during their critical sex-differentiating period in the womb. (See above under "Sex Differences from Conception Onward")

Towards Sexual Maturity

Because of these erotic differences in desire and mate choice, there are misunderstandings between the sexes. Women liberators caricature female's dislike of women seeming to be "*just* sexual objects" for men. Women want to be liked by men for their minds and their activities because women see these qualities as more important than the physical body. Women value certain qualities of men's minds (that may lead to wealth and social dominance) and companionship more than men's bodies. This is their value judgment influenced by their biology. Because they see the world through their mindset, they thus want to be desired more so for their minds, abilities and companionship than for their bodies. Because men appear to value a woman's form first before he enjoys her feminine mind, abilities and companionship, it seems to women that men do not like them as well as women like men. It is all a matter of how each sex values and perceives the other sex. Because men highly value women's appearance, they also want females to highly value their masculine body. Because women highly value men's mind, abilities, and social activity, women also want males to highly value them in the same way. But sexual maturity is when each sex learns to appreciate the others' differences and values. And it is through this kind of maturity that more harmony and understanding will occur between the sexes.

Perceptual Distractibility

Males' concentration is stronger and less distractible than females.[57] This comes from biological reasons having to do with males' higher levels of androgens than females[58], as well as how his mind is "wired." This is why males are more disturbed when interrupted when they are concentrating on something that interests them like football games, news programs, or the newspaper. This may be why wives or girlfriends are less likely to interrupt them.[59] She learns through trial and error that it keeps the peace not to interrupt. Conversely, this is probably why males verbally more often interrupt females in mixed groups,[60] for the females are less likely to be disturbed by such action. The androgenic influence on the focusing of attention and on the persistence and continuation of activities once initiated,[61] is the biological underlying reason for the perceptual distractibility difference.

Maternal Behavior

There is very sound proof that maternal instinct has to do with whether the individual has a male or female "brain." This maternal behavior is wired prenatally. Females who were accidentally masculinized prenatally through androgenic action in the brain show little maternal behavior. John Money wrote:

> Androgenized girls differed from their matched controls in preferred toys of childhood. They were indifferent to dolls, or openly neglectful of them. They turned instead to cars, trucks, and guns, and other toys that traditionally belong to boys.
>
> "Lack of interest in dolls later became a lack of interest in infants..... Some of the girls in this group distinctly disliked

> handling little babies.... By contrast, many of the control girls rated high enthusiasm for little children....
>
> The majority of the fetally androgenized girls subordinated marriage to career.... Among the control girls, the emphasis was in favor of marriage over nonmarital career.[62]

Conversely, girls with Turner's syndrome show even more maternal and feminine behavior than their control group of normal females.[63] Turner's syndrome is a chromosome condition where no androgenic hormones affect the brain prenatally.

Money also reported an experiment in *male* rats where they were induced to behave in a maternal manner through treatment with female hormones after they had been demasculinized in their critical period.[64] [Note: Rat's *critical period* is after birth, while humans *critical period* is in the 7th/8th week of pregnancy.]

Moreover, the male transsexual, who seemingly lives, works, thinks, and makes love like a woman, nevertheless, has little maternal urge for the newborn.[65] This is so because even though he acts outwardly like a female, he is still a *biological* male. And as a biological male, he has a biological male brain. He has a biological male brain because of the prenatal androgenic action on his developing brain.

All of these examples give us proof of a biological nature to maternal behavior. When androgens reach the brain of a prenatal infant at the critical stage, it sets up a non-maternal brain organization in the individual. When androgens do *not* reach the brain, prenatally, the infant develops a maternal brain organization that manifests itself in maternal behavior in later life, unless she is adversely conditioned.

Brizendine in her, *The Female Brain*, gives a popularized review of the biological evidence for the way hormones induce maternal behavior in her chapter called, "The Mommy Brain."

> Deeply buried in my genetic code were triggers for basic mothering behavior that were primed by the hormones of pregnancy, activated by childbirth, and reinforced by close, physical contact with my child. (p. 95)
>
> The mommy brain is switched on right at birth by a cascade of oxytocin. (p. 101)
>
> For the human mother, the lovely smells of her newborn's head, skin [...] will become chemically imprinted on her brain

— and she will be able to pick out her own baby's smell above all others with about 90 percent accuracy. This goes for her baby's cry and body movements, too.... Within hours to days, overwhelming protectiveness may seize her. Maternal aggression sets in. (p. 102)

As long as you're in continuous physical contact with the child, your brain will release oxytocin and form the circuits needed to make and maintain the mommy brain. (p.103)

She has about 60 endnotes in her book for this chapter on the "mommy brain" as she calls it. The chapter is a good read for understanding the influence of hormones on mothering and why men just cannot be as good at mothering as women. The author's goal was to get ordinary people to read her book so they can better understand the lastest scientific information on sexual differences. She is a medical doctor and is the founder of the Women's and Teen Girls' Mood and Hormone Clinic and was previously on the faculty at the Harvard Medical School and is a graduate of the Yale University School of Medicine and the U of Cal at Berkeley, in neurobiology.

Androgenic and Estrogenic Influence

Secondary effects. Besides the androgenic action of differentiating the human race into males and females, which creates the functional difference in each sex, which is important in understanding the difference in behavior between the sexes, the action of the androgens, indirectly and directly, also create secondary effects. Some secondary effects are the extra body hair on males, the coarser skin of males, the shorter life span of males,[66] the larger size of males, the lower basal pulse rate of males, the higher energy of males, and so forth. These differences play an important part in each sex's identifying with its own sex and differentiating its sex from the other.

Physical aggression and submission is also influenced by the levels of androgens and estrogens.[67] High levels of androgens help to produce high physical aggression. Estrogens have a depressive affect on physical aggressive behavior. Because of males' higher levels of androgens, they are more frequently physically aggressive than females. Male's aggression sometimes takes on antisocial manifestations (destructive or disobedient behavior). Their expression of aggression may "involve large muscles and the bringing together of bodies in blunt contact."[68] Because of females' lower level of androgens and higher level of estrogens, they are less as physically aggressive than males. Females' aggression takes on prosocial manifestation (e.g. spanking a disobedient doll to make the doll behave in a socially proper manner). Their expression of aggression does "not bring bodies, animate or inanimate, into contact nearly so much" as males.[69] Girls are more likely to have dolls hide from mother's calls, or be disobedient than to have the dolls physically destroy something.

Aggression *per se* is not exclusive of males, for females show in many cases just as much verbal aggression.[70] This verbal aggression of females is in the form of verbal bites, interpersonal rejection, getting others to intervene for them, tattling against someone else, and so forth.[71] Male's aggression, outside of their verbal aggression, takes the form in childhood of hitting, kicking, wrestling, and destroying toys through rough treatment. As males grow older they use less actual physical aggression and more likely use verbal aggression, backed up by their physical appearance.

Ambition and Drive. Also closely related to aggression are ambition and drive.[72] These three all have to do with the *vigor of behavior*. Males are more physically vigorous due to the influence of androgens in their brain development, and due to the everyday influence of androgens on activity level. This higher internal biological vigor is one reason men need less *social* reinforcement than women to drive to the completion of tasks. Of course, this drive is culturally relative. Psychological problems, economic problems, and social problems can slow down men's innate drive, while positive environmental factors can increase women's drive. **But as this work indicates, biology wins in the long run because it is continuous, internal, and persistent versus the environmental factors more temporary and inconsistent influence.** Moreover, the society that tries to overcome biological influences causes internal conflicts that weaken the people so acted upon.

Homeostasis

Females have more variability in body functions; males' body functions have greater stability. Women's various biochemical levels vary during their menstrual cycle. The main cause of this female cyclic pattern is known to be controlled by the hypothalamic brain cycle.[73] The hypothalamus of males is noncyclic in nature. Since the hypothalamus is important in regulating vital functions, including sex, then this sex difference in the brain "is pertinent, even though indirectly, to matters of psychosexual [gender identity] and behavioral dimorphism."[74] Besides sexual functions, the hypothalamus is important in temperature regulation, emotion, motivation, and other activities.[75] Thus, the cyclic nature of this subcortical brain structure is a biological reason for women's fluctuation in temperature, emotion,[76] and other activities.

Hormones and Growth

Improvements in dietary quality intake tend to increase the magnitude of physical sex differences in size from birth to adulthood. But even in impoverished nutritional areas, women are generally smaller than men. This is caused by hormonal action.

Androgens. Androgens help to increase size because they promote the synthesis of proteins from fat and amino acids, and they facilitate the retention of calcium, phosphorus, nitrogen and potassium.[77] All this assists in muscle and bone growth and repair. Since men have a higher ratio of androgens to estrogens, they have larger muscles and bones than women. Because of the males' higher levels of androgens, they develop muscular strength and size easier than females. This is why when women weightlift they do not develop the mass of muscles that men do when they weightlift, except those who take male hormonal supplements.

Estrogens. In contrast, estrogens help to break down proteins and make it more likely that dietary fats are metabolized and stored with fattish tissue. Estrogens also facilitate weight gain through their special ability to retain water. Women's change in weight during their menstrual cycle is caused by the changing levels of estrogens. Women's higher levels of estrogen verses androgens make it almost impossible for them to compete against men in most physical activities where size, strength, and other physical qualities are important. Yet females' earlier puberty gives them some physical advantage over many males in the 11th to 14th years.

Maturational Rate

At birth girls are better-developed infants than boys. The female newborn is from 1 month to 1½ months ahead of the male neonate in developmental acceleration. She fully matures physically by the age of 20 years versus age 24 for males. Females' bone structures are ossified from 8 to 27 months earlier than males. Girls reach puberty earlier than boys, and their maximum growth rate is 2⅓ years earlier than boys. Females stop growing before males. Boys continue to grow rapidly two or more years after the growth rate of girls decreases markedly at about 15.

Although the maturation rate varies according to what is measured, generally, females' biological functions are ahead of males in development at birth until about 15 years of age. Therefore, it has been put forth that the average male's poorer results than female's in school grades and various IQ tests and subtests during their earlier years may have something to do with the growth rate differences between the sexes.[78]

Developmental Timetables

Since young males are biologically lagging behind in development in comparison to young females, then of course, they should not do as well on average on IQ tests and school grades. In fact, "since the more mature girls would be expected to be ahead of the less mature boys in all behavioral manifestation, the absence of observed sex differences at birth might therefore conceal actually present sex differences in favor of boys."[79] But even though boys nearly catch-up to girls academically in late high school, which makes sense if physical and mental maturation go together, there are other environmental factors one must take into consideration.

Although girls at birth are four to six weeks ahead of boys in development, they, like the boys, have just been born and thus have no environmental experiences. Therefore, one could not say that the girls' intelligence at birth was four to six weeks ahead of the boys, just because she is biologically that far ahead, for the simple reason she has had no experience with her environment. We should remember it is the interaction of environmental factors and biological development that produces mental ability. Yet, as we explained before, the main cause of behavior is biology.

Eleanor Maccoby, a Professor of Psychology at Stanford University, disagrees with the idea that mental growth and physical growth are somehow related. Let's analyze her arguments to see if they invalidate the contention in this book that biology is the main cause of behavior, and hence the main cause of growth in mental ability.

Maccoby, in her article, "Sex Difference in Intellectual Functioning,"[80] reviews the theory of parallel developmental timetables between biological maturation rate and intellectual growth rate. She then quickly discounts this theory: "But Bayley[81] has shown that the rate of intellectual growth is unrelated to the rate of physical growth if one scores both in terms of percent of mature growth attained. Hence it does not appear that there is any single developmental timetable controlling both physical and mental growth."(p.39) This argument, in itself, does not negate the fact that biology is the main cause of behavior as we related before. After all, mentally limited children are so because they are *biologically* limited. And other arguments using "superior" biological individuals would also back-up our reasoning here: That is, biology and environment interact, but biology limits and causes behavior more so than the external elements.

Maccoby, then, does not invalidate our argument. What her use of Bayley's article seems to do is to negate the theory of parallel growth timetables between biology and intellect. Yet an examination of Nancy Bayley's work invalidates Maccoby's reasoning.

Maccoby says that the rates of intellectual and physical growth are unrelated "if one scores both in terms of the percent of mature growth attained." But Maccoby fails to tell us this: (1) Bayley arbitrarily sets age 21 as the time of 100% maturity in intelligence even though she notes that in a group she tested the participants were still gaining in intelligence at age 25;[82] (2) Bayley set the time of 100% maturity in physical growth at about 16 years; (3) Using the children's 16th year intelligence scores as l00% maturity level, there was a surprisingly high positive correlation between growth in intelligence and physical stature (p. 70ff); (4) There was only a negative correlation between biological growth and intellectual growth when the 21-year intelligence scores were used as the full maturity level. What these four items tell us together is that the rate of intellectual growth and physical growth is related up to 16 years of age when the majority of physical growth stops. And the reason intelligence growth continues after the 16th year is because it is produced by the interaction of the mature developed brain with continued experience. If children never learned after the 16th year, their intelligence would stagnate. Therefore, contrary to what Maccoby asserts, there is good evidence that the development timetables of biology and mental ability are related. One reason men begin to equal women after the 16th year is that at that time they have finally caught up to females in biological maturity.

Males and females have a different biological organization of the brain [83][15], from which they perceive the world and act upon the world differently. Because of this biological difference, males are superior in certain activities much as females are superior in certain activities. Just because the male brain is different in some biological ways doesn't mean it is superior or inferior. Both brains are complementary systems.

> Maccoby also added in her criticism of the developmental timetable theory, that: "It is difficult to see, for example, why maturational factors should produce greater differences between the sexes in spatial than verbal performances. Nor why a fast developing organism [female] should show different kinds of relationships between intellectual functions and personality traits than a slow-developing organism [male]."[84]

The answer to this is that besides the different rates of growth, the sexes have different biological organizational factors in the brain, different functions and structures which influence behavior, and different rates of growth for various parts of each sex's body. Bayley's article, which Maccoby cites, relates to this latter fact: "...not only do structure and function develop and become differentiated from each other, but also they do this at different rates. *These differences in timing occur for different aspects of a single organism, as well as between different organisms.*"[85] (my emphasis)

Homosexuality

It is possible for a normal-looking male to have a more or less feminized "brain" because his brain was not masculinized enough in the womb through androgenic action.[86] Not only this, but it is possible for a normal-looking female to have a more or less masculinized "brain" because her brain was masculinized in the womb by her mother taking hormonal substances (medication) at the critical stage in her development.[86] Although this is possible and has happened in the past, it is believed that such individuals are few in number. But there is actually no way of knowing just how many girls have androgenic masculinized brains because of prenatal intrusion of androgenic-like hormones or various drugs that cause the same kind of masculinized effects.[87] Therefore, because of this precariousness in fetal development, it is possible in a some cases to have a human with the genitals of one sex but with the apparent masculinity/femininity characteristics of the other sex.[88] This possibility may be one explanation for *a few* homosexuals and transsexuals, but appearing to be more feminine than the average male does not make one a homosexual. There are numerous feminine-appearing men and masculine-appearing women who are heterosexual. Of course, there may be other reasons also for homosexuality (the "other mind"),[96] but in this chapter we are only examining biological or environmental/socialization reasons. The scientific fact is that **there is no real creditable evidence to show that there is a genetic trait or gene for homosexuality**.[97] There have been three "studies" which were once often cited to suggest an inherited homosexual trait (Simon LeVay, *Science*, [1991], 253, pp. 1034-1037; J.M. Baily and R.C. Pillard in Dec., 1991, *Arch Gen Psychiatry*, pp.1089-96; Dean Hamer study in July 1993, *Science*.), but all three have been scientifically discredited.[98] In other words, no one is genetically *born* a homosexual. Homosexuals make up 1-3% of the general population.[99],[100] [90][91]

Animal Research's Relevance

In this chapter we have listed some of the effects sex hormones have on humans. Some of these effects are from research on mammals besides humans. According to radical feminist thinking, animal studies do not prove anything against the feministic principle which says that the environment is mostly the cause of behavior.[92] We say that animal studies on the effects of sex hormones have much to do with humans. Scientists have long used other animals for research when testing new drugs for man. How do the radical feminists think the oral birth-control pill was tested at first? It was tested through experiments on animals. The reason research on animals can be valuable to humans is because we have many things in common with them. We do have things that are not in common.

---

[96] "Other-Mind" in the *New Mind Papers* : http://becomingone.org/NM20.html

[97] NE & BK Whitehead, *My Genes Made Me Do It! Homosexuality and the Scientific Evidence*, 6th Edition, 2020: https://mygenes.co.nz/download.html

[98] Ibid.

[99] U.S. Census Bureau, Statistical Abstract of the United States: 2012, "Births, Deaths, Marriages, and Divorces," Table 97

[100] National Health Statistics Report (Number 77, July 15, 2014), "Sexual Orientation and Health Among U.S. Adults: National Health Interview Survey"

4. Sex Differences and their Implications

For example, we can speak a complex language; they don't.[101] But, in research between animals and us, we can be reasonably certain that if a sex hormone affects a mammalian animal in one way, it will also more than likely affect us in a very similar way. We can be sure because these animals also have gonads, reproductive tracts, internal and external genitalia, androgens and estrogens, estrous or menstrual cycles, and so forth. The nearer the animal research concerns similar attributes of mankind, the surer we can be that the ascertained effects will also be effects in mankind. We have to wonder if radical feminists would eat poisonous foods just because they were only proven poisonous in rats? We wonder if radical feminists would eat cancer-causing foods just because the food was only proven to cause cancer in rats?

One feminist, Naomi Weisstein, in a paper called, "Psychology Constructs The Female," tried to rationalize animal studies away by using an analogy between what animals do not have (speaking ability) and what humans do have (speaking ability): "Following this logic, it would be as reasonable to conclude that it is quite useless to teach human infants to speak since it has been tried with chimpanzees and it does not work."[93] Using this kind of analogy to disprove the worthiness of animal studies is reckless. It is fundamental knowledge that an analogy between two things is as good as the two things are alike. To make the analogy Weisstein uses is poor reasoning. Animal studies are pertinent as long as what is studied is similar in both the animals and humans.

Since the sex hormones of mammals are similar, since there are male and female mammalian animals and mammalian humans, since female animals give birth like female humans, since chemical reactions in both are similar, since both have similar brain structures, [94] then because of these similarities and others, animal studies can be reasonably used to ascertain potential hormonal effects in mankind. David A. Hamburg of Stanford University, Department of Psychiatry, put it this way. "We chose chimpanzees for our study because they are probably man's closest living relative. There are

---

[101] Research into great ape language has involved teaching chimpanzees, bonobos, gorillas, and orangutans to communicate with human beings and with each other using sign language, physical tokens, and lexigrams; see Yerkish. Some primatologists argue that the primates' use of these tools indicates their ability to use "language", although this is not consistent with some definitions of that term. https://en.wikipedia.org/wiki/Great_ape_language

many similarities between man and chimpanzee in the number and form of chromosomes, in blood proteins, in immune responses, in DNA, in brain structure and behaviour." [95]

Physical Sex Differences

Now we shall list and examine various worldwide gender differences. We will begin with the obvious biological differences. One should remember that we will be speaking in general terms for many of the items that follow. There may be some exceptions in some cases, but generally the following sex differences are true for the average male and female. The information on these physical differences was taken from the books listed at the beginning of the chapter. It must be noted that because of the great increase in obesity among men and women since I first wrote this book, some of the average dimensions listed may not be applicable today.

Genitalia

Males have penises and testes; females have vaginas, ovaries and wombs.

Form

Males and females differ in form. Their curves, angles and shapes are different, generally and relatively. Females generally have a more roundish look because their subcutaneous fat covers and hides their muscles. Men do not have women's characteristic layers of fat beneath their skin, and therefore their appearance is more roughish because their muscle tissues show through their skin more so than females.

Head Features

Women's eyes are set further apart than men's. Women's eyebrows are lighter than men's in appearance. Looking toward the front, women's faces are rounder, broader than men's. Looking from the top down on the head, women's heads are rounder, while men's heads are longer from front to back.

Breasts and Shoulders

Women have developed breasts with larger nipples and areolas than men. Women's shoulders are narrower, rounded, and more sloping than men's.

Angle of Arms and Legs

The angle of women's thighs and lower legs gives a "knock-knee" effect to females, while men's form a straight line. Also women's arms form a bent "carrying angle" at the elbow, while men's "carrying angle" is straight.

Hips and Legs

Looking towards the front, women's hips are wider than men's, and their hips have a more roundish curve than men's. Women's legs have a conical shape, while men's legs have a cylindrical look.

Hands and Feet

Women's hands and feet are relatively smaller, narrower, and more delicate looking than men's.

Hair

Women do not have noticeable hair like men on their chest, arms, legs and other bodily areas. Women's pubic hair is formed like a triangle pointing down; men's pubic hair forms a triangle pointing up. Women do not lose head hair like many men do in old age.

Dimensions

Women are generally smaller and more delicate than men. Physical sex dimensional differences vary in different regions of the world by race and quality of food intake. The following apply generally for those of western European descent:

Height

Females are shorter than men by 1 to 2% from birth to 11 years. From about the 11th to 14th years females are taller (to 2%) than men because their puberty is sooner than men by about 2 to 3 years. At about 15 years males overtake females until at 20 years women are generally 10% shorter than men.

Weight

Females from birth to 11 years are generally lighter than men by about 5%. In the 11th to 15th year girls weigh more until at the 14th year females are 5% heavier than males. In the 15th or 16th year boys regain their weight advantage, and by 20 women are generally 20% lighter than men.

Strength

Generally, males are stronger at all ages than females. The males' strength advantage increases after their puberty. This is caused through the action of males' androgens which facilitate the synthesis of proteins. And, as we know, protein is the food of muscle. Men are generally 50% to 60% stronger than women. This strength difference is one cause of the sex division in labor.

Vital Capacity and Muscular Tension

Vital Capacity

Vital capacity is the volume of air that one can expel from one's lungs after a maximum inhalation. By the 6th year boys' vital capacity is 7% higher than girls. This advantage of males increases to 10 to 12% by the 10th year, and to about 35% at age 20. Also the "vital index," the ratio between vital capacity and weight, is higher for males at all ages measured. This is important because higher physical activity and sustained energy output require greater oxygen consumption. This is one reason for males' greater motor activity than females, and one reason why boys seem more restless and vigorous than girls.

Muscular Tension

Males are more physically restless than females. They need more gross muscular activity. Because of males' greater muscular size, potential energy output, vital capacity, and androgenic influences toward physical activity, their muscles exhibit greater tension than females if not exercised.

Nutrition

Males have a greater quantity need for food intake. They need higher levels of protein, calcium, potassium, calories, and so forth than females even when males are the same weight as females. This is so because males have a higher Basal Metabolic Rate than females. Men use greater amounts of fuel than women because their body runs at a higher level of activity.

Biological Defects

Males have greater quantities of biological defects than females:

> more males are color blind;
>
> more males are born stillborn;
>
> male infants have higher rates of mortality and morbidity;
>
> males are more susceptible to many diseases;
>
> males grow and mature physically slower than females;
>
> among males there are more learning and behavior disorders;
>
> a higher percentage of males are mentally defective;
>
> and males develop their verbal abilities later than females. [96]

These biological defects are felt by many to have something to do with males' XY chromosomes and other genetic and hormonal factors.

These sex differences are important aspects that separate men and women. These sex differences and others exist because of the differences in hormones between the sexes.

Biology Limits and Causes Behavior

Absolute and Relative Limits

There can be no doubt that biology limits behavior. There are many obvious examples:

1. Fish are limited to water environments;
2. Mankind is limited to oxygen environments;
3. All living animals except mankind are biologically excluded from verbal complex communication.

All three of these examples are *absolute* limits of biology. The biologies of fish, mankind, and animals would have to be altered in order for them to respectively breathe air, live in outer space, or talk. And, of course, if these creatures are biologically changed, they would not be fish, or men, or animals, but something different. Besides absolute limits of biology there are also **relative** limits:

1. There are biological limits to how fast men or women can run;
2. There are biological limits to how much men or women can learn (brain organization and life-spans are factors).

In regard to males and females, the absolute biological limits between them are first the cells (XY v. XX), the functional difference in their brains, the reproductive ones and their related hormones that go with these reproductive differences.

The relative biological limits between the sexes are such factors as strength, spatial ability, verbal ability, etc. and these also have to do with genes, DNA and hormones.

Now the radical feminists believe there are only *minor* differences between the sexes. In a sense the differences between males and females are minor compared to, let's say, man and dog. But we know there is at least one *major* difference that sets the foundation for behavior differences between the sexes. The fact that women bear babies limits females, as the fact that men don't bear babies limits males in various ways. Some tell us that biology can be overcome. Some futuristic ones even suggest that test-tube babies (grown totally outside the womb) are the ultimate answer to allow women the "freedom" of men. They tell us we can set up child care centers to help set women "free" to work outside the home. They tell us we can use machines to compensate for differences in strength. But they overlook biology.

4. Sex Differences and their Implications

Science is far from growing babies outside the womb. And even if it were possible, it is psychologically improbable that mankind would accept it, except maybe a few radical feminists. And if it were possible, and if some mad group or leader forced it upon us, the resulting children would be mother-deprived children,[102] and in the end this would weaken the society and eventually destroy it.

As we have mentioned and will show in greater depth in Chapter 5, childcare centers do not work well, because the children so reared are inferior when compared to those with a closer, more constant mother-child relationship. A nation that enforces mass childcare centers will soon find that their children are inferior compared to other nations, and will eventually be seriously weakened by the experiment. Some radical feminists might argue, saying, institutional childrearing isn't all that bad, but the evidence is against such radical feminists.[see Chapter 5] Furthermore, child care centers are economically expensive. Women would likely staff them, and therefore women would not really be "freed" from children in order to advance their careers.

There are also many economic reasons why machines can't be used in mass to wipe out every need for male-like strength in heavy jobs. It just can't be done in the near future due to the world's economic, pollution, energy, and other problems. To try and overcome biology is expensive. The only reason some women with children can still work in their professions is because they underpay the females that care for their children. It is much easier to accept biology. It is much easier to give status to traditional women's work and/or supply more family friendly part-time jobs (through laws) than it is to shove women with young children into full-time outside-the-home jobs just so they can feel they are as good as men.[103] Sure women can do much, if not most of men's work, but it is economically reckless and stressful on women (and their children), since women must also bear the children if humans are to continue. The radicals have set up a false path for women to follow, so they too can "be just like men." It has not and will not lead to paradise, but only to disappointment and frustration.

---

[102] see the motherhood Chapter 5 of this book

[103] There is a problem with self-worth and confidence in women and that too is caused in part by biological differences: *The Confidence Code*, by Katty Kay and Claire Shipman, 2014.

Biology is the Main Cause of Behavior

Yet not only does biology *limit* behavior, it also *causes* behavior. In fact, biology is the main cause of behavior.

Biology is the main cause of sexually differentiated behavior in mankind. Many feminists believe the main cause is environmental. Most authorities on sexual differences tell us that the only true answer for the cause of sexual behavior is the interaction of biology and environment. We are not saying that the interaction or interplay of biology and environment doesn't cause some sexual-differentiated behavior, for some differences are caused by the interaction. But we are saying that the **main** cause of most sexual behavior is because of the biological nature of mankind.

Food. First, let's give another example of biology causing behavior. What causes a person to eat? Is it biology? Is it the person's environment or culture? Or is it the interaction of both biological and environmental reasons? Does someone eat merely because when he was growing up his parents always ate, and he learned from them to eat? No, of course not, he eats because his body (biology) needs nourishment to survive and because his parents taught him eating food from his environment was the best way to satisfy his need. If his body was self-sufficient, he wouldn't need food. And without the food, he wouldn't have anything to eat. Therefore, the interaction of both his biology and the fact that his environment has foodstuff to eat causes him to eat. But the real reason and main cause for a man to eat is biology — his biological nature is not self-sufficient. He biologically needs to eat. If his biology was different, and if food was available, the food itself would not cause him to eat. A person eats because there is a biological need for him to eat. His biology is the *main* cause of his eating behavior.

Clothes. Next, let's look at an example of a relative biological cause of behavior. Take for an example a man who leaves a house of 70 degrees fahrenheit into the outdoors of 30 degrees fahrenheit. After entering the cooler environment, he puts on warmer clothes. What causes the man to put on the warmer clothes? Is it biology? Is it the environment? Is it the culture? Or is it the interaction of biological and environmental reasons?

Environmental reasons do play a part. The weather is cold (environment). The man is living in an age where men who are cold usually put on warmer clothes (culture). And yes, the interaction of biological (his body's reaction to the cold) and environmental reasons is the cause of the man putting on warmer clothes. But what was the *main* reason or cause of his behavior? It wasn't the environment, for other creatures (polar bears) would not have to put on clothes to keep warm in cold weather. The low temperature is merely a secondary cause. But the real and *main* cause is that the man has biological limits — his body can only tolerate a certain degree of coldness.

Biology is the Main Cause of <u>Sex</u> Behavior

In regard to sexual-related behavior, biology is also the main cause of the behavior. The obvious, of course, is birth. Women give birth because they have the biological equipment to give birth and because it is needed for the survival of the human race. Because of this, nature gives women the desire to give birth (psychological hunger through her hormones and female brain) much like nature gives her the desire (hunger) to eat, and for much the same reason — survival. The main cause is not because females have "traditionally" given birth (culture), but because nature predestinates women to give birth.

Women usually feed and care for their children not merely because it is culturally taught that they should do this, but because of biological reasons. They have breasts with milk; men do not. Breast milk is the best food for infants.[98] Women are more nurturing towards children than men (proven through comparative culture studies). Women are of less strength than males, and have other different qualities than men. Thus through various processes a division of labor was established: women mostly care for children; men mostly do not care for the children. The previous example of Israel's <u>Kibbutz</u> is a good example of this process.

Women are usually less physically aggressive than men.[99] This has been proven repeatedly as hormonally caused. Males have a higher ratio of androgens to estrogens than females, and males' greater size and strength in comparison to females are some of the causes for the males' greater physical aggression.[100] Cultural conditioning may influence males to be less aggressive, but in reality biology cannot be easily handled in this way because these sexually-differentiated tendencies that come from within are as consistent as his ratio of androgens in his body.

Bio v. Culture Forces

Although biological drives in mankind have less effect on them than other creatures, and hence they are more culturally malleable,

[101] the innate drives of mankind will nevertheless **win out because innate biological drives are continuous while cultural forces are arbitrary, discontinuous, and sporadic**. I'll repeat, innate biological drives are continuous while cultural forces are arbitrary, discontinuous, and sporadic.

Nations Against Biology. Groups or nations that allow its males and females to behave according to their biological tendencies will be more economically stronger and more emotionally stronger than nations that go against the nature of the sexes, for the nation that goes against the biologically directed drives of the sexes, will have to spend too much time and money to condition their population to act against their inclinations. This causes internal reaction and detrimental emotional effects. The nation that spends great amounts of time and money on training men to be maternal and women to be physically aggressive because of some naive ideological theory, is not doing what is biologically the easiest for the sexes, and thus is at a great disadvantage compared to other nations which do not try to fight biology. Cultures that fight biology waste time, energy and money while emotionally disturbing its members. When we see relativity consistent differing behaviors between males and females, we see this mainly because of the biological differences, not culture.

Review

From this chapter, we see that the brain is sexually differentiated prenatally, and that it does influence behavior of the sexes. Although the environment is another influence on behavior, the steady all pervasive influence of biology dictates and limits the behavior of the sexes.

As the sexes age from childhood their behavior sexually differentiates to noticeable degrees because:
1. they continue to biologically differentiate, especially at puberty;
2. they have greater access to different stimuli;
3. they perceive things differently;
4. they identify more with their own adult sex[104] (which acts differently from the other because of functional differences, strength differences, mental differences, etc.);

---

[104] If the child has only one parent, or two "parents" of the same sex, this is where problems occur. See "Fatherless Children" and "Parenting by Homosexuals."

4. Sex Differences and their Implications

5. they are at times differently treated by parents and society;
6. their innate biological differences influence their behavior.

Consequently, there are sex differences in play activity, reading interest, media interest, values, life goals, occupational motives, vocational interest, job satisfaction, school grades, affiliation needs, "achievement" needs, creativity, memory, problem solving, mechanical ability, spatial ability, verbal ability, mathematical ability, and so forth.[97]

The realists understand the biological reality, as well as understand that each sex is not homogenized: *some* in each sex may have more of a certain quality than the average person in the opposite sex. The realists take all facts into consideration when forming their dogma as opposed to the radical feminists who base their dogma more on myth than on reality.

References for Chapter 4

[1: ch 4] pp. 492-493 in 196 of the Bibliography list
[2: ch 4] 4; 61; 191; 59; 58
129; 185; 151; 18
[3: ch 4] 113
[4: ch 4] 120; 59; 86; 9
[5: ch 4] 47; 113
[6: ch 4] 149
[7: ch 4] p. 42 in 86
[8: ch 4] chap. 4 in 120
[9: ch 4] 120
[10: ch 4] 9
[11: ch 4] chap. 2 in 9
[12: ch 4] 120
[13: ch 4] 120; 59
[14: ch 4] 120
[15: ch 4] 113 pp. 39-49
[16: ch 4] 113 pp. 9-20; 28a
[17: ch 4] p. 460 in 4; see 185; 200; 180
[18: ch 4] 180 p. 270
[19: ch 4] 101; 4
[20: ch 4] 191
[21: ch 4] 83a
[22: ch 4] 86
[23: ch 4] 61
[24: ch 4] 61; 9
[25: ch 4] p. 349 in 103

[26: ch 4] pp. 35-37, 203 in **103**
[27: ch 4] pp. 12-13 in **103**
[28: ch 4] **191**; **101**
[29: ch 4] **191**; **101**; **169**
[30: ch 4] **21**; **154**; **169**
[31: ch 4] **83a**
[32: ch 4] **61**; **169**; **21**; **104**
[33: ch 4] **32**
[34: ch 4] **198**
[35: ch 4] p. 128 in **86**
[36: ch 4] **117**
[37: ch 4] chapters 2 & 3 in **103**
[38: ch 4] **102**
[39: ch 4] **205**
[39a: ch 4], p. 202 in **61**
[40: ch 4] **205**
[41: ch 4] **47a**
[42: ch 4] **72** — this was bibliography reference # 167 in **61**
[43: ch 4] **182** — this was bibliography reference # 401 in **61**
[44: ch 4] **94** — this was bibliography reference # 229 in **61**
[45: ch 4] p. 207 in **61**
[46: ch 4] **61**
[47: ch 4] **31**; **8**; **166**; **48**
[48: ch 4] **51**; **82**; **115**
[49: ch 4] **118**; see also **120** & **115**
[50: ch 4] **163**; **153**
[51: ch 4] p. 251 in **120**
[52: ch 4] **120**
[53: ch 4] p. 1398 in **115** & see **56**
[54: ch 4] **71**
[55: ch 4] p. 150 in **71**
[56: ch 4] p. 252 in **120**
[57: ch 4] **118**; p. 264 in **18**
[58: ch 4] pp. 118-119 in **86**
[59: ch 4] **215**
[60: ch 4] **216**
[61: ch 4] **6**
[62: ch 4] **120**
[63: ch 4] pp. 107ff in **120**
[64: ch 4] p. 84 in **120**
[65: ch 4] **118**
[66: ch 4] **74**
[67: ch 4] **208**; **209**; **86**; **146**; **140**; **152**
[68: ch 4] pp. 137-143 in **156**
[69: ch 4] pp. 137-136 in **156**

4. Sex Differences and their Implications

[70: ch 4] **61**; **86**; **9**; **156**
[71: ch 4] **9**
[72: ch 4] **86**
[73: ch 4] chap. 4 in **120**
[74: ch 4] p. 55 in **120**
[75: ch 4] p. 727 in **149**
[76: ch 4] chap. 3 in **9**
[77: ch 4] **86**
[78: ch 4] **61**
[79: ch 4] **61**
[80: ch 4] **101**
[81: ch 4] **17**
[82: ch 4] pp. 68ff in **17**
[83: ch 4] **120**; **59**; **9**; **86**
[84: ch 4] p. 40 in **101**
[85: ch 4] **17**
[86: ch 4] **120**
[87: ch 4] p. 86 in **120**
[88: ch 4] pp. 58-59 in **120**
[89: ch 4] **58**
[90: ch 4] **178a** p. 196
[91: ch 4] **211** p. 219
[92: ch 4] **201**
[93: ch 4] p. 218 in **196**
[94: ch 4] pp. 236ff in **120**
[95: ch 4] **73a** p. 19
[96: ch 4] **165**; **61**
[97: ch 4] **61**
[98: ch 4] **131**
[99: ch 4] pp. 323ff in **133**
[100: ch 4] pp. 108ff in **86**; **61**; **152**; **120**; etc.
[101: ch 4] pp. 292ff in **18**

5. Motherhood and Maternal Deprivation

Motherhood or the position of being a mother is systematically attacked by the *radical* women liberators (see Chapter 2). To these feminists motherhood keeps women "trapped" in the home with the children. Motherhood is enslaving, degrading, and a thankless self-sacrifice that pays no money and receives no status, according to the radical feminists and their associates. Like the male chauvinists, the women liberators think that traditional male occupations are greater in value than motherhood. Babies, cooking, and other aspects of motherhood are degrading, according to the propaganda. To be fulfilled is to go outside the home into the "mainstream" of society.

The radical feminists are so bourgeois that they seem ignorant that most jobs outside the home are tedious and lacking in any kind of creativity and expression of individuality. The only reason it seems to be better to the radical feminists is because they fall for the male ego-building tactic of considering anything that males do more important than anything females do. It is almost as if men had an innate need to differentiate themselves from women. This may have something to do with males being reared in their early years almost exclusively under their mother's control[1], or because now in today's industrialized and computerized society, they have less biological protection in their male occupations. Women are biologically protected in motherhood (men cannot have babies), except in the eyes of the 2020s trans movement. Or the male's attitude may come from a combination of several cultural and biological reasons. But whatever the reason, most chauvinistic males think what they do is more important than whatever females do. And some/many women fall for this value judgment of men.

Radical feminists often fall for this male's value judgment because they have identified with the ideas and ideals of men. But no matter what the propaganda, motherhood is a vital and important aspect of human existence. Children with smart, full-time[105] motherly rearing (with a father in the home) are nourished with the important qualities only a full-time mother can bring to childrearing.

---

[105] especially in early childhood; mother's with part-time jobs may be classified as full-time mothers

Maternal Deprivation

In a 1952 monograph, *Maternal Care and Mental Health*,[2] the author J. Bowlby, concluded that young children's prolonged deprivation of maternal care may have grave and far-reaching effects on the character of the children. Bowlby's paper was criticized for its supposed over-simplification.

In 1962 another monograph appeared by a half-dozen authors, *Deprivation of Maternal Care*.[3] In this book the evidence up to 1962 was reviewed by Mary Ainsworth:

> "An examination of the evidence should leave no doubt that maternal deprivation in infancy and early childhood indeed has an adverse effect on development both during the deprivation experience and for a longer or shorter time after deprivation is relieved, and that severe deprivation experiences *can* lead in some cases to grave effects that resist reversal."[4]

This work dealt with the deprivation matter in a more complex manner than Bowlby's work, yet came to similar conclusions.

In a 1972 book, *Determinants of Behavioral Development*,[5] by Leon Yarrow, Chapter 4 examined the maternal deprivation problem. The research of the 1960's was included in this work. Similar conclusions about the adverse effects of maternal deprivation were noted in this work. Leon Yarrow mentions that the matter of maternal deprivation is quite complex and can't be simplified to one form of deprivation. When trying to measure maternal deprivation such factors as duration, nature, and severity of maternal deprivation must be taken into account. Furthermore other biological variables of the child, and the nature of the child's total environment must also be taken into account.

How Does Maternal Deprivation Occur?

From Mary Ainsworth's review we quote:

> Severe maternal deprivation is now known to occur under the following diverse conditions:
>
> (a) when an infant or young child is separated from his mother or permanent mother-substitute and cared for in an institution where he receives insufficient maternal care;

(b) when a young child undergoes a series of separations from his mother and/or substitute mother-figures to whom he had formed attachments;

(c) when an infant or young child is given grossly insufficient maternal care by his own mother or permanent substitute mother and has no adequate mothering from other figures to mitigate the insufficiency of mother-child interaction....

There is much evidence that the discontinuity of relations brought about through separation from the mother-figure or surrogate-figure (after an attachment has been established and before the child is old enough to maintain his attachment securely throughout a period of absence) is in itself disturbing to the child regardless of the extent to which the separation ushers in a period of deprivation or insufficiency of interpersonal interaction....

It is now clear that deprivation occurring without physical separation can in fact be as pathogenic as deprivation occurring with separation.[6]

What is Wrong With Institutional Child Care?

From Ainsworth's review:

As David & Appell[7] and others have pointed out, a multiplicity of mother-figures tends to obviate sufficient adult-child interaction. In most institutions where each child, in the course of a day, has many caretakers, each adult has partial responsibility for many children. Under these circumstances two factors combine to give insufficiency of interaction: the adult does not have time to give much stimulation to any one child; the adult cannot be sensitive to the behavior of any one child, so that he does not respond to many of the child's potentially [sic] social signals. Rheigngold[8] has demonstrated that this pattern of caretaking makes for decreased social responsiveness in the infant under twelve months of age, even in an institution where the total amount of care given to each child is not grossly and obviously insufficient. If deprivation of this kind is extreme or prolonged into the second year of life and beyond, the result can include the grave effects which, by now, are well known.[9]

5. MOTHERHOOD AND MATERNAL DEPRIVATION

Examples of Institutional Child Care

From a paper on Soviet women by Mark Field and Karin Flynn we quote:

> If the hypothesis of the effects of maternal deprivation is a valid one not only for Western cultures, then the Soviet scheme should be a source of important information for the planners for institutionalized societies. Some evidence has come to light that Soviet children raised in a nonfamilial atmosphere, invariably perform more poorly in their school work than those who come from families, even though these families might be economically deprived. One of the main reasons for this is that the children in institutions do not have a single adult person with whom they can identify and establish a continuous relationship. *In one Soviet school, for example, children could not recognize themselves in group pictures because someone had forgotten to hang mirrors.* In another such institution, the personnel were so busy in routine tasks that they had hardly any time to converse with their charges[10].[11] (my emphasis)

From a paper reviewing Josef Langmeier's 20 years of experience with Czechoslovakia's institutionalized children we quote:

> This organization of institutional child care represents a unique natural experiment. It has lasted almost 20 years and allows a follow-up of the children who have grown up under uniform and rather well-defined living conditions. Even though these conditions vary according to age group, they can be characterized in a general way as follows:
>
> 1. ***Lack of stimulation***. This is marked in infants' homes where there is reduced tactile, kinesthetic, and visual stimulation with a simultaneous excess of acoustic noise.
>
> 2. ***Lack of conditions for social learning***. The treatment of all children as a group makes it possible to train the children — particularly in the first years of life — in basic personal habits, but not in the modulations that are required in most instances of social interaction.
>
> 3. ***Lack of conditions for developing specific, lasting, and intimate social ties***. The changing of nurses, the transferring of children from one group to another, and the removal from one institution to another make it almost impossible to

establish lasting and strong ties between the child and particular nurse....

4. *Lack of conditions for autonomous action, for responsible decision-making, and for finding one's own identity.* As a general rule, children in institutions possess nothing of their own, have no privacy, rarely are allowed to play or work alone, and are not encouraged in achievement efforts.[12]

Effects of Maternal Deprivation on Children?

From Mary Ainsworth's review of the evidence in 1962 we quote:

> Maternal deprivation has a differential effect on different processes; most vulnerable seem to be certain intellectual processes, especially language and abstraction, and certain aspects of personality, most especially the ability to establish and maintain deep and meaningful interpersonal relations, but also the ability to control impulse in the interest of long-range goals....It seems likely that discontinuity of relations has its chief effect on the capacity for affectional ties, especially in instances where separation from mother-figures is repeated.'[13]

From a review by Langmeier in 1972 we see the effect of maternal deprivation on children's personalities:

> (1) In many instances of social interaction, they do not modulate their behavior in socially required ways. This is due to "insufficient amount, complexity, or variation in stimulation" of maternal deprived children.
>
> (2) They lack structure in their perception of their environment. To them, their environment is too chaotic, and beyond any meaningful order. This is because many maternal deprived children are not taught to deal with any environment except an all too commonplace or stereotyped one.
>
> (3) They lack the ability to form specific and lasting attachments to other persons. This is because they do not have any specific object in early childhood to be personally attached to. Usually, this "object" in early childhood is one's mother.

5. Motherhood and Maternal Deprivation

(4) They lack individualism or autonomy in their behavior. This is due to their lack of private ownership, their lack of learned motivation to be an individual, and to the fact that in the institutions they can rarely play or work alone.[14]

Harlow's Motherless Effect

Indifference of Unmothered Children Towards Motherhood

Even though childbearing is a great achievement, it is downgraded by the so-called enlightened feminists. It is probably downgraded because males can't bear children, and thus their downgrading of childbearing. Women may feel intuitively the worth of bearing a child, but she will be in need of reassurance by her man. Thus a manifestation of the identity of radical feminists with the values of chauvinistic males is shown in their indifference towards motherhood. Another aspect of radical feminists' indifference towards motherhood is that some of them were "motherless." Some radicals manifest the Harlow's motherless monkey effect. In Harlow's own words he writes about the non-maternal behavior of the unmothered mothers' behavior:

> **Month after month female monkeys that never knew a real mother themselves become mothers — helpless, hopeless, heartless mothers devoid, or almost devoid, of any maternal feeling.**[15]

Here is more detail about unmothered mothers in Harlow's own words:

> As we have already indicated, four of our rhesus females that had not been reared by real monkey mothers eventually become mothers, and we refer to them as our motherless, unmothered, or nonmothered mothers. It should be remembered that they had suffered more than real-mother deprivation since they had never been given any opportunity to associate with other infants, and thus were prevented from forming normal infant-infant affectional attachments....
>
> After the birth of her baby, the first of these unmothered mothers ignored the infant and sat relatively motionless at one side of the living cage, staring fixedly into space hour after hour. If a human observer approached and threatened

either the baby or the mother, there was no counter threat. During the first week postpartum this animal was the most catatonic monkey we have ever seen. It was necessary to remove the baby and to feed it at appropriate intervals, and even these procedures elicited no response from the mother.

As the infant matured and became mobile, it made continual, desperate attempts to effect maternal contact. These attempts were consistently repulsed by the mother. She would brush the baby away or restrain it by pushing the baby's face to the woven-wire floor....

The next two mothers to give birth were even more abusive to their infants than was the first. They ignored their offspring except when repelling the infants' advances. From time to time these mothers engaged in unprovoked aggression against the helpless neonates....The fourth mother resembled the first, being indifferent more often than abusive....

...However, none of the unmothered mothers ever showed more than transient indications of any positive affection for their young. In other words, in so far as infant-mother attachments were concerned, it was a case of the baby adopting the mother, not the mother adopting the baby.[16]

Indifference to Sex. Not only were the motherless monkeys indifferent and abusive towards their offspring, but before their pregnancy they were quite indifferent towards sexual relations with male monkeys. Even in their "heat" period, they were sexually indifferent toward the males. It took systematical plans by Harlow and his associates to get the motherless monkeys pregnant.[17]

Harlow protege Stephen J. Suomi has continued to experiment with rhesus monkeys at the University of Wisconsin Primate Laboratory.... Infant monkeys reared exclusively with peers, rather than a mother, from birth, interact later with other monkeys 'but not healthily,' says Suomi. Such monkeys are abnormally fearful and 'cling to each other most of the time,' he reports.

Even brief maternal separations have lasting effects later in life, Suomi says. Monkeys removed from their mothers for just two hours a week during the first eight months of life demonstrated effects such as an unnatural fear of the

5. Motherhood and Maternal Deprivation

environment for as long as two years later. (Joel Greenberg, p. 139, *Science News*, Aug 27, 1977)

Radical feminists who were reared by surrogate mothers, because their middleclass mothers were too busy with their careers, act some- what like Harlow's motherless monkeys. One such feminist, Marya Mannes, wrote:

> There was nothing strange whatever to me in the fact that my mother, a woman, spent much of each day practicing or giving lessons, that she often went off on tour with my father, and that she couldn't boil an egg. She didn't have to. In the early part of the twentieth century even people of very modest means had cooks and nurses [what? — she projects her sheltered middleclass state here], and it was taken equally for granted by my brother and myself that if our mother was away, the cook, the nurse, or the great-aunt who lived with us would take care of us....
>
> ...My mother was an exceptional woman.... [the] servants and relatives made it possible for her to maintain a professional life and a domestic life....
>
> It was therefore quite natural that I should grow up believing that all was possible for a girl or woman....
>
> But at no time, then or since, did I throw away a sense of fierce independence as a human being and the desire to attain distinction in terms of mind and spirit and expression....I did not even, at that early age, want children.[18]

Marya Mannes goes on in her paper, "The Problems of Creative Women," to call for alternate roles for women besides motherhood. She, like her mother and other radical feminists, identifies with chauvinistic male values, and probably helped by her "motherless" childhood, has come to further disassociate from motherhood.

By this summation on maternal deprivation, we can see the importance of mothering.[19] We need to know more about other aspects of motherhood like childbearing and childrearing.

Childbearing and Childrearing

Motherhood is intimately involved with children. To become a mother, a female must bear a child. This involves pregnancy which lasts for nine months. For nine months a woman carries in her womb a developing human. A mother actually creates a child with her body through the processes of her body, with help from her husband's sperm. And not just any form of life is created, but the highest observable one. A human life is much more complex than any mathematician's formula. It is the ultimate observable creation.

Having a child involves great responsibility. Although, by observing western society today, we can readily see it is not held as a great responsibility. But a human being is a complex creature that needs much attention, especially in the early years, and throughout the first 18 to 20 years. Having a child not only involves the proper rearing of a complex and warmth-seeking individual, but it involves 20 or more years of a mother's life. Therefore, one should not have a child for the wrong motives, for to do so may be quite unfair to the child and the mother. What are some of the wrong motives for having a child? And how will these motives adversely affect the child?

Wrong Motives for Childbearing

From an article by Anna and Arnold Silverman, "The Wrong Reasons for Motherhood,"[20] we learn that some women think that by having a child they will gain additional recognition: "My child will be somebody." And through her child obtaining some sort of status, the mother believes she will gain some additional status among her acquaintances. This is using children. Such mothers, the authors say, sometimes push their children towards various achievement goals, not for the sake of the child, but for their own status in the community. They have a tendency to push their children too much, and help to cause emotional problems in their children.

According to the authors, some women have children because they think they will save their marriage. They believe the marriage will be cemented by the child or children. The problem with this motive for having children is that the mother is again using the child. Such mothers have children because they feel insecure about their marriage lasting, and they think a child or another child will help to save or secure their marriage. If the marriage fails such mothers may give their child or children too much affection (forming an almost psychological symbiosis), or blaming the break-up of the marriage on the child or children.

Other wrong reasons for having children besides the reasons already given are:

    (1) "Having a child will prove I'm a sexually mature woman."

    (2) "I had an unhappy childhood. My child won't."

(3) "Having a child will make me an adult."
(4) "Having children will give me something to do in my old age."
(5) "Having a child will please my parents."
(6) "My religion expects married women to have children."
(7) "Without children you're an outsider."

If any of these above reasons is the *overriding* reason a woman has a child, she is having a child for mostly selfish reasons. The problem with having a child for selfish motives is that the child will more than likely suffer some detrimental consequence in his or her psychological make-up.

To paraphrase the authors, the right reasons for having children should include altruistic reasons (to bring another life into the world that will be reared into a healthy individual), as well as reasons that pertain to the mother (a child will fulfill her biological functions and desires). Anna and Arnold Silverman believe that altruistic reasons for having a child should be the main reason for having a child. We disagree. Altruistic reasons *and* reasons that pertain to the mother should be on the same level. Some so-called selfish reasons for having children are not necessarily bad reasons. They only become bad reasons when the children's right to be constructive individuals is thereby denied. The mother-child relationship is a two-way street. Both mother and child have rights in such a relationship. Both mother and child deserve to get something good out of such a relationship, if both put something good into it.

Nursing

Motherhood not only consists of bearing a child, it also involves nursing a child and rearing the child into adulthood.

The best way to nurse a child is to breastfeed.[21] Breast milk is especially designed for babies:

(a) it has the right balance of nutrients;

(b) it is uncontaminated — you know it is not from some diseased cow, or from a contaminated container;

(c) it is at the right temperature and has no harmful bacteria;

(d) it is inexpensive;

(e) the colostrum, secreted with the milk of the breasts for several days after birth, is rich in antibodies that fight off infections;

(f) babies fed with breast milk are less likely to develop allergies and less susceptible to infection than babies who are fed cow's milk;

(g) it is all an infant needs for the first 6 months in the way of nutrition;

(h) and besides the economics and nourishment value of breastfeeding, there is that all important emotional and loving-warmth benefit for both mother and child.

One of the main problems with Harlow's motherless monkeys was that they lacked the warm and soft skin contact with their mothers. This is one reason they were so indifferent and abusive to their offspring.

Therefore, considering the above, breast-feeding is an important part of motherhood, and should not be exchanged for the bottle and its inferior cow's milk. Cow's milk is for calves; human breast milk is for human infants. Because of the subtle and open bias against breast-feeding, some women actually need training to feed their child with their breast. Such organizations as Le Leche League International should be consulted for information by women who wish to breast-feed their infants. Because of the subtle and open indoctrination against breast-feeding, because of the increasing numbers of working mothers, and because of other reasons in the mid-1970s only about 20% of the Western world's women breast-fed their babies at all and only 6% breast-fed their babies for more than 6 weeks.[22] This rate increased dynamically in the 1980s where as many as 25% in the United States breast-fed their babies up to 5-6 months, and 60% at least breast-fed their babies in the hospital.[106] In 2013, in the USA, breast-feeding for the first six months of an infant's life is up to about 49%.[107] In some countries, like Russia, "most Russian babies are nursed, often up to a year," but mothers are kept from their infants for the first 24 hours.[23] The problem with this Russian tradition of keeping the infants away from their mothers for

---

[106] Peter Weiss, *San Jose Mercury News*, August 18, 1992, p. 1E

[107] Breastfeeding Report Card, United States, 2013, CDC report— http://www.cdc.gov/breastfeeding/pdf/2013breastfeedingreportcard.pdf

the first 24 hours is that the child misses one day's supply of colostrum, which is only secreted with the breast milk of the mother for a few days after the child is born. Furthermore, for those mothers who give their children unrestricted breast-feeding for a long period after birth, they find that mood cycles that accompany ovulation and menstruation are absent for a long period after birth since their menstruation cycle is absent. In a study by D.S. Matthews of 374 Nigerian women who nursed 18 to 22 months in an unrestricted manner, it was found that menstrual periods resumed an average of 16 months after delivery versus the average of 55 to 59 days for non-nursing women.[24] Considering everything, breast-feeding is a great activity that a mother can do for her child and herself. But because of the push to get women into the marketplace — even mothers of small children — babies in the Western world are being deprived of the breast.

Childrearing

In addition to bearing and nursing children, motherhood includes rearing the children properly. Rearing children involves teaching children. Since the human mind is quite complex and flexible, when compared to other animals, such teaching of children can be involved. There is nothing boring about properly training children. It is a full-time job and can be quite intriguing and challenging when done in a proper manner.

Radical feminists think rearing children can be quite boring and non-fulfilling. This is so with radical feminists because they have identified with men and their ideas and ideals. To men childrearing may seem boring or unfulfilling merely because their interest is elsewhere since their biological destiny is elsewhere. This doesn't mean men should not take interest in childrearing. It only means they do not take up as much time as women with childrearing due to biological and cultural reasons. The radical feminists, because they identify with chauvinistic men, are inclined to devalue the occupation of rearing their children, and for status purposes they spend too much of their time with other outside interests. These "mothers" are more interested in radical feminism, or social groups and clubs, or their church, or their school, or even their job than their children. But children are much more important than some job programming a computer that bills us, or that computes some esoteric formula. Children can be more rewarding than many other aspects of life, especially for women, because motherhood is their biological destiny.

How Should Children Be Reared?

Children should be reared to become healthy and constructive individuals. To do this they must be reared with the ability to cope with their future environment, not some utopian environment dreamed up in the minds of radical feminists or other unrealistic dreamers. If an utopian environment comes to earth, then (and only then) should our children be reared to function in such an environment. But as of now, and as our history has shown, there are no, nor have their been any utopian environments that have lasted for any number of years. We live in a competitive world. Our children should be taught to handle themselves properly in this world. They should not be sheltered from it, only to be pushed into it at some arbitrary age like 18 or 21 years.

In order for a child to handle himself in the world, he must be given steadily more freedom of action as he ages. In his early years, a child may have to be guided in almost everything he does. But in his later years, he should be given increasing responsibility and freedom, so that at the age of 18 years or so, he will be able to handle himself in the world as it is.

The parents should rear their child realistically. They should recognize the biological nature of their children. They should rear their girls to be women and mothers first, but knowing the realities of modern society, also empowering and enabling them with the abilities to obtain a job outside the home environment. They should rear their boys to be men and fathers, first, as well as enabling them to obtain skills to work in the modern world. They should guide their children to know their limits and potentials. In order to know their children's limits and potentials, they must spend time with their children and take interest in their destines. Both husband and wife should rear their children as a team in harmony. The parents should not pass on all their responsibility to the school system or day care centers. The most important education a child receives is from their parents,[25] for it is his early family experience that sets the foundation on which he later learns.[26]

Children's Education. The whole question of the state taking the traditional rights of parents over their children must be examined closely. Is it better for the state to teach our children to read and write or is it better for parents to do this task? One of the reasons the state takes away children from parents at six years old and demands that they go to school is because of the certain percentage of parents

5. Motherhood and Maternal Deprivation

who would not give their children any instruction. But what right does the state have in taking away responsibilities from parents just because of a minority? Isn't it better for parents to teach their children to read and write, and to teach them about morality? Some will differ and say parents aren't equipped to teach their children. To answer this, one only has to look at the way the so-called teaching profession has failed. Of course all parents are not equipped or knowledgeable enough to give their children all the various aspects of an education. In these cases though parents could then send their children to school only for the certain educational subjects that they lack knowledge in. And of course the state could always give tests to the children to see if the parents are actually teaching them to read and write.

Parents should have the first right to teach their children. Only if the parents fail in this duty should the state step in and put children in school. Motherhood would be a more challenging profession if the state had not taken the children away from the parents at six years, for in effect this is what the state has done. After six years of age our children are indoctrinated by the state in the subjects that it wishes to teach. Parents have little say as to what is to be taught their children. Just try and change some aspect of your school system if you don't believe this. And if the state had not taken away many of the responsibilities of parents, we would not have the radical feminists and other women complaining about being bored at home. So far the state has not taken away the children any earlier than six years although there is some talk of the state putting children in school as early as three years of age. Therefore since naturally mothers care more about their children than teachers in most cases, then mothers still have great opportunities up to six years of age to influence their children in positive ways. Mothers should make every attempt to be at home with their children as much as possible during these early years.

Considering everything, motherhood will be the prime goal of most women in the world as long as the human race reproduces itself. And because of motherhood and sex differences, there will be sex roles. And because of sex roles, there will be differences in behavior between the sexes since each role makes different demands and has different environments. Furthermore, other relative sex differences, such as muscular strength and brain organization, will also make for different behavior between the sexes.

As far as occupations outside the home, they have been and will be secondary to motherhood in nations that are actually reproducing themselves. Nations that are not reproducing themselves will see others immigrate to their nation that are reproducing themselves. The so-called women liberators try to reverse this. To them a profession or job outside of being a mother should be a woman's prime goal while the tasks of motherhood should be hired out to some day care center. But because of the female biological nature (her womb, breasts, hormones, etc.) and the economic and social realities of this world, most women in the world will seek, at least subconsciously, motherhood first, and a profession second. Some women will combine both motherhood and an occupation because of economic necessity or because of the desire for a better standard of living. But in this case the children will be deprived to the degree that their parents don't give them the needed attention.

We cannot cover all the aspects of motherhood in this paper. It deserves a book-length document. In this book we have listed various biological factors of females that set them off from males. Women's biological functions are needed in order to continue our human race. Their verbal fluency is needed in teaching children to talk, etc. Their more interpersonal qualities are needed for children's development. Their soft skin texture and breasts are needed for children's emotional and biological growth development. All of these factors and others give importance to motherhood. When women properly perform this needed occupation, they build the foundations of society. When they do not properly perform motherhood, the results are a detrimental to society.

Parenting by Homosexuals

Due to a 2015 ruling by the Supreme Court of the United States, homosexuals in all states will be able to marry and "parent" children. This was a tragic ruling that left biology, history, reason, and nature out of their ruling and allowed the pushers of anarchy to win in court. We believe and the best scientific evidence suggest this will lead to more troubled children at a much higher rate than children of divorced parents. The Family Research Institute, the Family Research Council and other groups have literature on this matter.[108] From their study on homosexual parents we quote:

Significant for both lesbian mothers (LM) and gay fathers (GF), with & without controls:

---

[108] http://familyresearchinst.org ; http://www.frc.org/homosexuality ; https://frc.org/issuebrief/homosexual-parent-study-summary-of-findings

5. MOTHERHOOD AND MATERNAL DEPRIVATION

Compared with children raised by their married biological parents ("intact biological family," or IBF), children of homosexual parents (LM and GF):
- Are *much* more likely to have received welfare growing up (IBF 17%; LM 69%; GF 57%)
- Have lower educational attainment (IBF 3.19; LM 2.39; GF 2.64)
- Report less safety and security in their family of origin (IBF 4.13; LM 3.12; GF 3.25)
- Report more ongoing "negative impact" from their family of origin (IBF 2.30; LM 3.13; GF 2.90)
- Are more likely to suffer from depression (IBF 1.83; LM 2.20; 2.18)
- Have been arrested more often (IBF 1.18; LM 1.68; GF 1.75)
- If they are female, have had more sexual partners-both male (IBF 2.79; LM 4.02; GF 5.92) *and* female (IBF 0.22; LM 1.04; GF 1.47)

Significant for lesbian mothers (LM) with & without controls
Compared to children from intact biological families (IBF), children of lesbian mothers:
- Are more likely to be currently cohabiting (IBF 9%; LM 24%)
- Are almost 4 times more likely to be currently on public assistance (IBF 10%; LM 38%)
- Are barely half as likely to be currently employed full-time (IBF 49%; LM 26%)
- Are more than 3 times more likely to be unemployed (IBF 8%; LM 28%)
- Are nearly 4 times more likely to identify as something *other than* entirely heterosexual (Identifies as entirely heterosexual: IBF 90%; LM 61%)
- Are 3 times as likely to have had an affair while married or cohabiting (IBF 13%; LM 40%)
- Are an astonishing *11 times more likely to have been "touched sexually by a parent or other adult caregiver"* in childhood (but not necessarily by the homosexual parent; IBF 2%; LM 23%)
- Are nearly 4 times as likely to have been "physically forced" to have sex against their will (at some time in their life, not necessarily in childhood; IBF 8%; LM 31%)
- Are more likely to have "attachment" problems related to the ability to depend on others (IBF 2.82; LM 3.43)
- Use marijuana more frequently (IBF 1.32; LM 1.84)
- Smoke more frequently (IBF 1.79; LM 2.76)

- Watch TV for long periods more frequently (IBF 3.01; LM 3.70)
- Have more often pled guilty to a non-minor offense (IBF 1.10; LM 1.36)

The shocking history of the radical homosexuality movement can be found in the book *The "Born Gay" Hoax* by Ryan Sorba. (http://www.massresistance.org/docs/gen/08a/born_gay_hoax/TheBornGayHoax.pdf)

References for Chapter 5

[1: ch 5] **35** of the Bibliography list
[2: ch 5] **26**
[3: ch 5] **210**
[4: ch 5] **2**
[5: ch 5] **123**
[6: ch 5] pp. 142, 143 in **2**
[7: ch 5] **39**; **40**
[8: ch 5] **143**
[9: ch 5] p. 145 in **2**
[10: ch 5] **6**
[11: ch 5] p. 229 in **54**
[12: ch 5] pp. 337-338 in **97**
[13: ch 5] p. 149 in **2**
[14: ch 5] pp. 332-336 in **97**
[15: ch 5] **77**
[16: ch 5] **78**
[17: ch 5] **78**
[18: ch 5] pp. 117-119 in **105**
[19: ch 5] see **27** & **150**
[20: ch 5] **164**
[21: ch 5] **131**; **25**; **63**; **145**
[22: ch 5] **28**
[23: ch 5] p. 144 in **168**
[24: ch 5] p. 70 in **131**
[25: ch 5] **179**
[26: ch 5] **125**

6. Answer to Radical Feminism

Feminism, Motherhood, Value, And Identity

As we've shown in this book, radical feminists have very negative attitudes towards motherhood and its various tasks. Motherhood is the position of being a mother. Some of these negative attitudes toward motherhood we have already stated previously. To the radical feminists *achievement* is not found in any aspects of motherhood. These women subsequently must go elsewhere to achieve or succeed. And that elsewhere is into the world of male-dominated status occupations. Women liberators and others think the housewife is "just a housewife." A mother is "just a mother." Having a child is "just bearing a child." And to the radical feminists the prenatal child is a parasite growing off of its mother and keeping its mother from "equality." To them an aborted baby is this a bunch of mush. Thus their holy grail is abortion. This attitude toward babies is one reason they support abortion because it frees women to be the *new* women — to be just like men. In short, the traditional work and functions of women are accounted by radical feminists and others as non-achievement and of low status. Why?

Many radical feminists have achievement mixed up with status. First in their mind achievement has little to do with motherhood. They are wrong and most non-indoctrinated women and men don't agree. But in the radical feminist mind, if you have status, you have achievement and vice versa. A female medical doctor is an achiever, or thought to be by American radical feminists. But in Russia where women once made up over 70% of the medical doctors, this profession was/is a low status one compared to American doctors. [1] Although, even this is even changing as more females become doctors. Why?

Why do women with Ph. D's with about the same salary and rank as their male colleagues feel discriminated against, and feel they have less status than their male colleagues?[2] Why do most, if not all, tasks, or women-dominated occupations seem to many to have a lower status than men's? Why does the same paper, when said to be written by a woman, have a lower value in men's as well as women's minds, than when said to by written by a man?[3]

Why do many women often overrate work done by males, and many men underrate work done by females? Why do women seek the status of some profession, enter and do well in that profession, and yet not reach comparable status with their male counterparts? Why do they paint themselves as "the other," the "second sex," the "lost sex," and so on? Why do women liberators in their writings manifest inferiority complexes? (But normal women don't really manifest this.) Why?

Because feminists, in their naivety, are following men's values and goals, *they* are inadvertently saying that men are the value-makers and that their biological makeup are of less value. They in affect are teaming up with the chauvinistic men. They are discounting their own inner nature, feelings and biological make-up and identifying with men's. This is a loosing adventure that has led to women being depressed and over medicated, and many are left without children or loving husband. But because of their march they have dragged many other non-feminist along with them who are also without children and loving husband.

To *Feminists* Men are The Value-Makers

If men went into the home and said, "this is status," then the radical feminist women would fight to get back into the home. Margaret Mead understood this phenomenon as she wrote in her book, *Male and Female*:

> In every known human society, the male's need for achievement can be recognized. Men may cook, or weave or dress dolls or hunt hummingbirds, but if such activities are appropriate occupations for men, then the whole society, men and women alike, votes them as important. When the same occupations are performed by women, they are regarded as less important. In a great number of human societies men's sureness of their sex role is tied up with their right, or ability, to practice some activity that women are not allowed to practice. Their maleness, in fact, has to be underwritten by preventing women from entering some field or performing some feat. Here may be found the relationship between maleness and pride; that is a need for prestige that will outstrip the prestige which is accorded to any woman.... cultures frequently phrase achievement as something that women do not or cannot do, rather than directly as something which men do well.[4]

The feminist, Dr. Cynthia Fuchs Epstein, has said about the same, but in a roundabout way:

> "The greater social desirability of a type of work, the less likely it is that women are identified with it. All societies seem to prefer men in the jobs most valued."[5]

Male Feminists. Parenthetically, as we said before many radical feminists are middle class or bourgeois, and thus, so are their husbands. Many middle class men think their male ways are better, and if their wife takes up beliefs which adhere to the superiority of male ways, then some, if not many, of the men will encourage their wives to pursue feminism. This type of reasoning also applies to some non-feminist men, no matter what class we are speaking about. Thus, some men think feminism is alright because it preaches what they value for themselves. They consequently think, through their one-sided reasoning, that women also value as they value. Hence they think it is good for women to strive to be like men through such movements as women's lib.

Even the old Equal Rights Amendment seemed to have been for the radical feminists a psychologically uplifting Amendment to make women in general *feel* less insecure about their worthiness. Adel T. Weaver, a former president of the National Association of Women Lawyers, in 1971, said the following before the House of Representatives in its committee on the Equal Rights Amendment:

> "Therefore, I urge you to enact the equal rights amendment because the American woman today really does not yet recognize her own value as a person and a human being, and until she is accorded the constitutional rights to equality under the law in every respect, she will not begin to free herself from the bondage of centuries of self-depreciation."[6]

Homosexuality Of Many Radical Feminists

Another proof that we can give, that many women regard men's values more than their own, or at least feel insecure about their own values, is the phenomenon of feminism. Many feministic values are men's values. Think a moment. Radical feminists think they will achieve "equality" if they become like men, and as one in a consciousness raising group said: "We're in here trying to be men, right?"[7] Right! Radical feminists are trying to be like men. They identify with men's values. This is one reason why they hate biological differences, for this would forever mean that they could not become like men. Radical liberators have made a mistake in identity. They want to be men. Some feminist leaders even identify with the male sexual desire. The desire of these homosexual radical feminists is for other females. The feminist Kate Millett has been quoted:

> "Of course I'm bisexual. We all are. This is the revolution. The women's movement has always had lesbians at its vanguard. Much of the running motor has been supplied by lesbians, even when they were in the closet. The lesbian is the archetypal feminist because she's not into men — she's the independent woman par excellence."[8]

Female homosexuals think of the women liberation movement as their chance to legitimize their peculiarity:

> "A vital relationship between lesbians and women's liberation is in their mutual interest in a time of changing relationships. Lesbians are the women who potentially can demonstrate life outside the male power structure that dominates marriage as well as every other aspect of our culture. Thus, the lesbian movement is not only related to women's liberation, it is at the very heart of it. The attitude toward lesbians is an indicator by which to measure the extent of women's actual liberation."[9]

This quote is from an article by Sidney Abbott and Barbara Love, who are lesbians according to *Esquire*, July 1973. Even the great mother of sex role malleability, Margaret Mead, is said to have been a homosexual. (see Chapter 3)

Feminism As A Form Of Male Chauvinism

Many radical feminists are thus like mirror images of male chauvinists. Women liberators' desires are males' desires. Radical feminists are not liberating women, they are entrapping them with well-sounding words into becoming like men. It should be just as pathetic or ludicrous to see women blindly running after men's values, as it is to see men in primitive cultures mimic child birth, as if they were giving birth — the so-called couvade ritual. The men who congratulate women in men's activity do so because they think these women have bettered themselves for taking on men's "superior" ways. These men are called friends of feminists, but they are really friends of male chauvinists who think that men's activities are superior.

Biological Protection. We will not attempt to go far into the question as to why men seem to be the value-makers to feminists, or whether it is due to nature, or nurture, or the interaction of both. One interesting possible answer to this question may be that it has something to do with the males' lack of biological protection of their role. For women's traditional role of motherhood, women do have biological protection — men cannot have babies. However, in a few societies today women do not have cultural protection of their role while men do have cultural protection. That is, in many societies men's roles are taught to both men and women as higher status roles than women's traditional role of motherhood. Thus, cultural indoctrination takes some security away from women's absolute biological protection of their true role of motherhood. Although women are biologically protected in their role, some are intellectually unsure that motherhood is a worthy occupation. The degree of this insecurity differs between women. Some women are quite secure with motherhood and think it, and what goes with it, is great. While others like the radical women liberators are not only insecure about motherhood, but they hate it, at least in their own writings. And men probably hold up their occupations and roles above women's in order to compensate for their lack of biological protection. But this doesn't tell us why many women fall for men's status game. Because this phenomenon is so prevailing, it must have something to do with biology. Women do give value to their own tasks, but in the mad feminist world of today they need to have *reassurance* from men they admire, or respect, or love as to their true worthiness. Even radical feminists are insecure about their ideals. They won't let or don't like men who enter into their feministic groups because of various fears, like thinking men may take control or use "male logic" and confuse them. Women liberators need reassurance from each other, or their husbands, or their boyfriends about their beliefs.

Answer To Feminism: Give Value Where Due

The radical feminist's "equality" will not make women happy just as life in the Israel's kibbutz did not make women happy.[109] Women have been the sole possessors of motherhood, and always will be if our human race is to continue. The way to make women happier is to make them more at peace with their true nature. Women are the life-givers of this planet. Without our mothers more of us would be like Harry F. Harlow's mother-deprived monkeys — cold, unloving, unfeeling, etc.[13] I get the feeling reading the radical feminist writings that they were more or less motherless, thus the explanation for their negative feelings toward motherhood. **The way to make women happier is to acknowledge the value of women's tasks**. Today, and since the 1970/80s, when most home economic classes where taken out of our schools, women in our country are not taught to be efficient mothers and wives, or taught to cook or about nutrition or about sewing, but are taught as if they were males — even though males and females have different functional directed biologies. This kind of educational environment sets up an ambivalent state for our women,[14] which isn't the healthiest state for happiness. From an article by Judith Bardwick and Elizabeth Douvan we quote:

> In spite of an equalitarian ideal in which the roles and contributions of the sexes are declared to be equal and complementary, both men and women esteem masculine qualities and achievements. Too many women evaluate their bodies, personality qualities, and roles as second-rate. When male criteria are the norms against which female performance, qualities, or goals are measured, then women are not equal. It is not only that the culture values masculine productivity more than feminine productivity. The essence of the derogation lies in the evolution of the masculine as the yardstick against which everything is measured. Since the sexes are different, women are defined as not men and that means not good, inferior. It is important to understand that women in this culture, as members of the culture, have internalized these self-destructive values....
>
> The culture generally rewards masculine endeavors....By these criteria, women have not produced equally. The contributions that most women make in the enhancement

---

[109] as shown in Chapter 3 of this book

and stabilization of relationships, their competence and self-discipline, their creation of life are less esteemed by men and women alike. It is disturbing to review the extent to which women perceive their responsibilities, goals, their very capacities, as inferior to males; it is similarly distressing to perceive how widespread this self-destructive self-concept is. Society values masculinity; when it is achieved it is rewarded. Society does not value femininity as highly; when it is achieved it is not as highly rewarded....

...Frustration is freely available to today's woman: if she participates fully in some professional capacity she runs the risk of being atypical and non feminine. If she does not achieve the traditional role she is likely to feel unfulfilled as a person, as a woman. If she undertakes both roles, she is likely to be uncertain about whether she is doing either very well. If she undertakes only the traditional role she is likely to feel frustrated as an able individual. Most difficult of all, the norms of what is acceptable, desirable, or preferable are no longer clear. As a result, it is more difficult to achieve a feminine (or masculine) identity, to achieve self-esteem because one is not certain when one has succeeded.[15]

Because men's values are the radical feminists values, and because the main street media propagate these values, the problem here is for men to recognize and understand women's value to the world, and to give women recognition, especially verbal — for most women respond to verbal compliments more so than material or symbolic or silent compliments.[16]

Sexes Need Sex-Differentiated Teaching

What women should do as teachers of children and homemakers is an important quality, but that quality is difficult to evaluate. Courses in school should be set up for males and females to understand the sex differences, and to perceive that the differences are of equivalent value to mankind. Parents should be taught the equivalent differences, and thus pass such knowledge on to their children.

Jean Grambs and Walter Waetjen,[17] wrote in *The National Elementary Principal* about the students' right of "Being Equally Different." They begin by mentioning that "the most unequal thing that happens in our schools is that unequals are treated as equals." They are referring to sex differences:

> ...our schools are sex-neutral institutions operating on the assumption that all persons are alike with respect to the ways in which they learn and achieve. We wish to make crystal clear our position that *it makes a significant difference whether the person we are teaching is a boy pupil or a girl pupil and that instructional provisions should be made accordingly.* (pp. 59-60)

Grambs and Weatjen mention the use and abuse of research. They mention that even though there are an abundance of papers which test and amplify sex differences, many major text books used in education ignore or "convey the idea that sex differences are not really important in learning." (p. 60) Grambs and Weatjen go on to explain how important the sex differences are, and why different methods should sometimes be used to teach each sex the same subject. The reason why different teaching methods should sometimes be used is because each sex conceptualizes things differently.

Feminized Grade School System. Grambs and Weatjen write of our feminine grade schools which misunderstand boys and which discriminate against boys in various ways, making it difficult for them to learn. The middle-class women teachers often do not understand the rough behavior (language, motion) of boys. Also the authors point out that "women literally do not know that they use words differently, structure space differently, perceive persons and reality differently from men." (p. 64) Thus, "men express continuous exasperation over women: 'They aren't logical. How can you follow what they mean?'" Women are not openly aware most of the time that men perceive their words from a different perspective, or vice versa. Therefore, it is very important for teachers to take into consideration sex differences in perception when they find male or female students doing poorly in certain subjects. The authors concluded that more men should teach in elementary school, or at least that the female teachers should be trained to understand the differences when assigning homework and giving tests.

Judith Bardwick, in her book, *Psychology of Women*, points out the feminized grade school system of this country and how it hinders young boys:

> Let us, for example, compare boys and girls when they enter the first grade. At this level, the school requires at least a

minimum of controlled behavior and an ability to memorize; first-grade children are expected to learn to read and write. For whatever it means (I'm not quite certain, but I read the phrase often enough), at this age girls are more mature than boys, perhaps by as much as a year. They are less overtly aggressive, they are less threatening to parents and teachers, they are well motivated to being "good" and loved, they are more socially oriented, and they are more verbal.

And the boys? The poor boys! Boys are less mature, and their impulsive and aggressive qualities threaten authority. Their higher impulse level requires greater inhibition in order for them to quietly concentrate and learn. There is more early stress on boys to conform to a sex model: thus the active boy has to restrain himself in order to do problem solving (under the threat of banishment to the principal's office) while the passive boy is rejected by his peers. The boy must meet early demands for inhibition in temperament, for the acquisition of verbal skills in which he does not excel, and for conformity to interpersonal norms. In addition, some of the areas in which he could excel, such as space manipulation and exploration, are not part of the problem-sets the school engages in. Thus the stress from the culture is greater on boys and the capacity for resolution is less. It is not surprising that the incidence of personality pathology in boys is much higher than in girls....

It seems to me that parents and teachers have to understand what is happening to the boys and cease insisting upon feminine standards of performance.[18]

Garai and Scheinfeld in their comprehensive paper on sex differences explain the different motivation for learning in each sex:

"Girls do well in subjects which draw parental approval or social approval; boys do well, not so much for social approval, then to conquer the task itself in which they are interested. Males find satisfaction in the task itself; women find satisfaction in the social recognition and praise for doing it well."[19]

Males are motivated to do things because they are oriented in "task achievement," while women are more oriented in "affiliation achievement." These sex differences in motivation and other sex differences in aggression and perception must be taken into consideration when teaching our children. But if the radical feminists had their way, our children would be taught as if the sexes have no differences in outlook, perception, biological destiny, and so forth. As it is now, our children are taught too much as if there were no sex differences between them.

What Should Be Done

We need a reverse revolution, a turnabout. It should be a political, religious and cultural movement to bring sex/gender roles back into harmony with biological reality. Women are the life givers — the child bearers. This was what was recognized and culturalized until industrialization and feminism pushed most women into working full-time outside the home, while giving little consideration to their necessary role as the life-givers. To restrict or deny this reality of motherhood is not only harmful to women, men, children and all society, but to the quality and satisfaction of life.

The false-feminism from the 1960-70s is not working well and will never work well because of the biological dictates of nature. Without the proper conditions for motherhood nations will *not* thrive because their children will be inferior compared to other cultures where the conditions for the family are better. Radical feminism was/is based only on myth that seeks to transform women into being just-like-men, while ignoring biology and the importance of the family. Many feminists do not even have children which robs them of the satisfaction of motherhood and leaves their pool of genes cut off from the world. No child will ever remember them or their careers. **But this is not the worse aspect of the old radical feminism:** it is not only the radicals that hurt themselves, most women under the old feminism are also likewise affected because many men in this environment don't want children and just want a wife that works so they can have more man-toys. The men and the media who are allowing and encouraging radical feminism are in effect dooming their own cultures by naively sabotaging the family, the home, the culture, the future, and the happiness of their own existence and their nation.[110]

Will the reverse revolution ever happen? Yes. Either after the collapse of the presently constructed world, or after the coming revolution that will end the dominance of the "other-mind" in the world. I await that new world.

---

[110] See chapter 4, under "Bio v. Culture Forces" for more details.

References for Chapter 6

[1: ch 6] p. 195 in **10** of Bibliography list and chapter 5 in **168**
[2: ch 6] p. 92 in **166**
[3: ch 6] **65**
[4: ch 6] p. 168 in **110**
[5: ch 6] **127**
[6: ch 6] p. 301, see p. 291 in **196**
[7: ch 6] p. 176 in **68**
[8: ch 6] p. 74 in **41**
[9: ch 6] p. 620 in **1**; note pp. 33-34 in **57**
[10: ch 6] p. 46, chap. 3 in **9**; p. 37 in **60**
[11: ch 6] chapter 4 in **9**
[12: ch 6] **189**
[13: ch 6] Chapter 7 in **78**
[14: ch 6] **11**
[15: ch 6] pp. 55, 56, 57 in **11**
[16: ch 6] **198**; pp. 192-193, chapter 6 in **61**
[17: ch 6] **69**
[18: ch 6] pp. 104-105 in **9**
[19: ch 6] pp. 228ff in **61**

7. Sex Fallacies & Sexual Maturity

There are various fallacies that males and females project on the opposite sex. Some we will list here. These mistaken ideas project the misunderstanding and prejudice between the sexes. When each sex becomes cognizant of these misperceptions it may lead to better relations between the sexes — a sexually mature attitude that negates chauvinism.

Gossip

Women gossip; men don't. This isn't true. The word "gossip" merely describes talk about other people. Women talk about *people* because women are interested and involved with people. Men also gossip, but they call it "shop talk."

Superstition

"Women are superstitious," so goes the fallacy. This statement is spoken by men to mean that women are highly superstitious while men are only slightly. Many women are superstitious. They are superstitious about things they are interested in, yet have little control over (who or when they will marry, etc.). But men are also superstitious about things they are interested in, yet have little control over (sports, gambling, etc.).

Courage

"Men are courageous, women aren't," so goes another fallacy. But both sexes are courageous in their own interests, functions, and destinies. Men are courageous in war; women are courageous in childbirth, etc.

Deception

"Men are such liars." "You can't trust a woman." Both sexes are deceptive. Men lie about certain things (business, wealth, etc.) that are important to them, or that they think are important to others, as women lie about certain things (people, their past, their shape [padded bras]) that are important to them, or that they think are important to others. Both lie at times to impress the other sex, to obtain something from the other sex, to keep something from the other sex, and so forth.

Vanity

One sex calls the other vain, and vice versa. Actually both are vain, but both are vain about what is important to each's own sex. A man is vain about his job, his car, his sport, his hobby; a woman is vain about her face, body, clothes, house, behavior, etc. To say that one sex's vanity is worse than the other is a biased value judgment.

Words and the Degradation of Females

A one-sided perceptual fallacy is manifested in the cries of radical feminists over the English language's abundance of words that degrade women. In an article for *Newsweek,* Elizabeth Peer tells us that the compilers of The Feminist English Dictionary, list over 400 words of a "sexist" nature, and:

There also are the disagreeable, specifically female epithets like "crone," "harpy," "hag," "shrew," "virago," "termagant," "harridan," and "slattern" for which male equivalents are rare. Even such putatively favorable words as "lady" are chauvinist putdowns, insists Todasco [editor of the dictionary].[1]

But a few weeks after Peer's article a male reader wrote *Newsweek* pointing out that the English language had as many degrading words for men: "I suggest consulting the following list and asking yourself which sex comes first to mind: jerk, creep, slob, bastard, beast, brute, louse, ass, pig, pimp...."[2] Females have degrading words for men, as males have them for women. Females also sometimes use men, as males sometimes use women. What one sex does to the other, the other does to that sex. Only the feminist selective perception fails to see this.

Achievement

"Men achieve; women don't," so goes the fallacy. The whole aspect with "achievement" is that it is measured in the male-dominated fields. When you read feminist papers, when you read psychologist's papers on achievement, they are speaking of achievement outside the home, outside motherhood. There is a built-in bias with the word "achievement." You can only achieve outside the home, or outside motherhood. This of course gives the advantage to males and is a great disadvantage to females. Women do achieve. Many are great achievers in motherhood and homemaking. The reasons many people don't recognize this is because: (a) men are, in some respects, apparently the value-makers; (b) men see their own activities as superior to women's; (c) to achieve for men means to achieve in their fields of endeavor; (d) since some women accept this male definition of achievement, they likewise think that achievement comes only in male-related activity, not necessarily in motherhood. The achievers in female-related activities are not seen as achievers because of the biased male driven idea of achievement. Women achieve as men do, but only a few feminists recognize it as achievement because of the special and biased way the word "achievement" is used. Women have the conflict between their biological desires for children and social definitions of achievement.

Practical Sense

"Men are more practical," goes the fallacy. However, both sexes are practical, but in different ways pertaining to their sex differences. For example, men say that Jane isn't practical, for she spends too much on clothes. But from her view-point such behavior is practical because: (a) she is single; (b) she wants to marry; (c) a good appearance attracts men; and (d) therefore good-looking clothes are practical for Jane. We see that what is practical for one sex, may *look* impractical to the other sex, but a fair inspection of the circumstances allows us to see the activity in a less biased manner.

Equality

"Equality between the sexes is a right," so goes the fallacy. We say equality — being treated *exactly* the same — is not a right, and furthermore is impossible since males and females are different. They are different in function, make-up, and behavior, as this book shows, and as other professional papers and books prove. No law, no ideology will make males and females the same. Males and females should be treated alike in many ways, but when it comes to sex differences, we should take them into consideration and treat each sex accordingly. We shouldn't use sex differences to enslave one or the other sex. We should be fair to each. *We should take into consideration sex differences when we make up social and legal laws, and reflect these sex differences in our laws.* Some of the "double standards" of society are understandable when one examines the sex differences between males and females.

Men use Women

In feminist writings we read how the "brutes" use and take advantage of women. These writings, by their one-sidedness, seem to say that only men use women, not vice versa. This isn't true. Men may take advantage of women, or use women sometimes, but sometimes women also take advantage of men and use men. Women, for various reasons (like "getting" their man), allow themselves to get pregnant on purpose. They use the sex desire of men to manipulate them. They threaten men with divorce if they don't get their way, and so forth. Some women use men only some of the time; some women use men most of the time. Some men use women some of the time; some men use women most of the time. Both sexes try to manipulate the other in various ways, at various times, and in various degrees.

Perception, Self-Centered

It is fallacy when one sex subconsciously thinks that the other sex thinks and perceives the world as they do. Males and females see things differently, as this book shows, and they give different value to what they see depending on how important the object being perceived is to them. Because of women's different biology and different environment, they see differently than men. This fact should be taken into consideration by males and females when they interact and deal with each other. *Much misunderstanding has occurred because males or females subconsciously think the other sex values and perceives as they do.*

Chauvinism

Men are male chauvinists, according to the feministic propaganda. And some/many men do act like their sex is the best. But women liberators seem to say only males are chauvinists. This again is a fallacy. Women are also chauvinists. Some women are even male chauvinists. As we have shown in this book the radical feminists are of the latter kind. Some women are female chauvinists. A female chauvinist is a woman who believes that typical female attributes are better, inherently, than typical male attributes. Such female chauvinists think that: (a) a beautiful female body is innately better than a handsome male body; (b) females are better innately because they seem more gentle, warm, and better-mannered; (c) romantic love and sex is innately better than male-lustful sex; and so forth. A female chauvinist thinks that male's sexual ways are disgusting. Their sexual ways are too "crude," too unromantic, too bodily oriented instead of spiritually or mentally oriented. All this merely projects their different needs and their different erotic perception. Female chauvinists think that the male's various manifestations of visual desire for the female form (through his stare, his nude magazines, etc.) are disgusting or are degrading to females because, in her hierarchy of values, visual desire of the opposite sex is much lower than her romantic desire for the opposite sex. This all has to do with the biological make-up of females as related previously in this work. Males are visually "turned-on" by females' form because of their biological organization and androgenic levels, not because they are sexist, inferior, or socialized improperly.

An example of a female's chauvinistic thinking is manifested in a short article by Marya Mannes in *Newsweek*. In her commentary Mannes says:

> "For the face remains the clue to the being, the persona. It is the light in the eye, the sound of the voice, the truth of words that tell man or woman what he or she needs to know in order to love or cherish: not the size of breast or length of leg. Beauty of flesh or body is indeed an *extra* gift."[3] (emphasis mine)

In her article Marya Mannes is critical of the new string swimming suits that were being worn by some young women at the beaches at the time she wrote her article. She tells us that the face, the voice, the words of an individual are more important than the "extra gift" of bodily form. This merely shows her typical female hierarchy of value. The male *chauvinists* would think that a woman's body was the most important aspect of a woman and that their ability, personality, charm were *extra* gifts. Both types of chauvinism lead to conflict, not harmony. A more mature aptitude would be to recognize each sexes biases (caused apparently by biological reasons) and make adjustments and compromises.

Sexual Maturity — The Goal

Instead of being either male or female chauvinists, each sex should learn about the various differences of the other's sexual desires and needs. Males should come to understand that females are not as interested in some things as they are, not because women don't like men, but because they like them in a different way. Females should come to understand that males are not as interested in some things as they are, not because men don't like women or respect them, but because they like them in a different way. Both sexes should learn to love each other in ways the other wants, needs, and desires. Both sexes should incorporate female and male ways of expressing love and desire when loving the other. **This is the true meaning of being sexually mature:** understanding the other sex's differences, accepting them as real, giving love as the other needs and wants. The two should learn to harmonize, combine, and become one as much as possible without radically mixing real sex roles that are based on biology. Let's recognize in this age that both sexes are self-centered in some ways, yet each sex should learn from the other, and harmonize with the other.

References of Chapter 7

[1: ch 7] **139** of Bibliography list

[2: ch 7] Sexist Epithets. In "Letters," *Newsweek*, Feb. 25, 1974, p. 7

[3: ch 7] **106**

8. Summary of Evidence and Issues

In this part we will summarize, discuss and give answers to various issues that radical feminists have brought to the attention of the public and government. Be sure to read again chapter 6 where we give the answer to radical feminism.

Each of the following discussions and their answers to feministic issues have been written to roughly stand alone. In this way one may quickly examine the issue of special interest to himself and find an answer to the issue. In conjunction with the other parts of this book, this part can be used in answering some specific feminist demands.

Conclusions on Sex Role Development

Sex roles are the way they are in this world because:

(1) *Biology limits and prepares each sex in different ways,* which in the long run has produced the traditional gender roles we have today. The traditional sex roles cut across all cultures because biology has dictated limits as to how far each sex can stray from their innate behavioral tendencies.

(2) *Biology limits and causes a bio-cultural pattern of behavior.* Since males and females are biologically dissimilar everywhere in the world, their biology limits and causes a universal bio-cultural pattern among mankind.

(3) *This bio-cultural pattern is incorporated within the thoughts and institutions of mankind.* The incorporation of behavioral patterns within the thoughts and institutions of mankind reinforces and magnifies the behavior patterns.

(4) *The parents reared in the universal bio-cultural pattern*, teach their children, directly or indirectly, consciously or subconsciously, to behave according to their proper gender roles.

(5) *The children accept their assigned gender roles* because their sex pertinent innate influences make it easy for them to accept their sex assigned roles.

(6) *If parents fail to rear their children in the traditional sex roles*, for ideological reasons or from ignorance, then the children are hampered in their development by: (a) their

opposing innate biological drives: (b) the opposing cultural pressures from those properly reared.

(7) ***If a whole society tries to go against the historical reality of traditional sex roles***, then that society helps to diminish itself because: (a) the society drives each individual against their innate biological drives (which weakens the individual); and (b) the society because of (a) begins to fall behind other societies in productivity/well being since too much energy is used to promote their ideology and to fight against their biology.

Socialization Theory Cannot Explain

As we have tried to show in this paper the socialization theory can't explain many aspects of sex differences. The following are some of the facts that socialization can't explain:

(1) It can't explain why vast majority of the world's people live within the hard traditional sex roles, and always have as far as records show.

(2) It can't explain why sex-differentiation is in the same direction everywhere: most males act in traditional masculine ways; most females act in traditional feminine ways.

(3) It can't explain the tomboyish behavior of the female androgenized hermaphrodites.

(4) It can't explain the "hyperfeminine" behavior of those with Turner's syndrome.

(5) It can't explain the sex differences in spatial ability or verbal ability.

(6) It can't explain the differences in the vigor of activity between boys and girls.

(7) It can't explain boys' more aggressive, assertive behavior, or their more asocial behavior in comparison to girls.

(8) It can't explain girls more affiliational needs in comparison to boys.

(9) It can't explain physical differences between the sexes in such aspects as strength, height, maturational rate, and so forth.

(10) It can't explain the male and female brains. (see Chapter 4)

Overlapping Rationale

Although females on the average are superior to males in some abilities, and although males on the average are superior to females at some other abilities, there are a certain few in each sex group that are better in the superior ability of the other group. For example, some females do better in higher mathematics than the average male. Conversely, some males do better than the average female in some verbal tasks. These are examples of ability overlapping between the female sex and the male sex. The professor of psychology, Anne Anastasi, puts it this way:

> ... any relationship found between group averages will not necessarily hold for individual cases. Even when one group excels another by a large and significant amount, individuals can be found in the "inferior" group who will surpass certain individuals in the "superior" group.[1]

Therefore, there is some overlapping between the sexes in talent, ability, and other biological factors. Jobs *per se* with heavy lifting should not be excluded for women just because most women can't do heavy work. If a woman can lift 70 pounds all day, and if she wants such a job, then she should have it according to the feminists. This sounds good, but many overdo this argument and make it seem like there is a great overlapping of differences. Although there is some overlapping in ability between the sexes, there are other aspects of the sexes that do not overlap:

> (1) All normal males have the XY chromosomes in each cell of their bodies.
>
> (2) All normal females have the XX chromosomes in each cell of their bodies.
>
> (3) All men have a higher ratio of androgens to estrogens in their bodies.
>
> (4) All women have a higher ratio of estrogens to androgens in their bodies.
>
> (5) All men have a noncyclic hypothalamic pattern in their brain.

(6) All women have a cyclic hypothalamic pattern in their brain.

(7) All men can't bear children or have a menstrual cycle.

(8) All normal women can bear children, and do have menstrual cycles.

(9) All men do not secrete milk.

(10) All normal women can secrete milk.

(11) All men have external genitals.

(12) All women have internal genitals.

(13) Any exception to (1) through (12) is very rare[111] and may because the individual is not a true male or female, but a hermaphrodite — a sex biologically mixed individual, or during gestation (and afterward) excessive hormones relating to the opposite sex were introduced into their systems at critical stages in development.

Because the cells of the sexes are different they react differently to the same amount of hormones.[2] Because of the protein synthesis action of androgens, because of the higher assertion action of androgens, and because of other effects of androgens, men will be stronger and more physically assertive than women. Because of the other androgenic effects, men will differ accordingly, even against women of the same size and intellectual capacity.

Because estrogens decrease fatty substances in the blood,[3] women have fewer heart attacks than men who have eaten the same amount of fatty foods. Because estrogens facilitate weight gain through its fat retention ability and water retention ability, women will carry more fat than men. Because of other estrogenic effects, women will differ from men of the same size and intellectual capacity.

Because of the differences in "brains," men and women will differentiate even if both groups were subjected to the same stimuli. Because of childbearing ability differences, men and women will differentiate even if both groups are subjected to the same stimuli. Because of the differences in genitals, men and women will differ in love making, identity, and eroticism.

---

[111] "By 1991, approximately 500 cases had appeared in literature." https://journals.sagepub.com/doi/10.2350/14-04-1466-PB.1

In order to eradicate these differences, one would need to change the biology of each sex. But even if women take doses of androgens, in order to become physically stronger, they must still give birth in order for the human race to survive. And in order to give birth to normal healthy children, women's hormonal levels must be normal, not abnormal. Other means of changing the biology of either the male or female sex would eventually lead to psychological and economical problems, if initiated. Add to this the high cost of trying to overcome biology, and one must ask — why fight it? The answer is that radical feminists are confused. They desperately want to feel achievement, but have chosen and identified with male achievement. Thus, they are biologically at a disadvantage, and therefore, have a psychological need to destroy what is in their way. Since sex differences are in their way, they call for sameness, or as they call it — equality.

To conclude then, there is some overlapping of ability. But we see that there are many biological differences between the sexes which lead to differences in behavior, interest, and ability. The overlapping argument is excessively emphasized by radical feminists in order to try and get their philosophy accepted in the face of obvious biological sex differences.

Teaching and Sex Differences

Various studies have emphasized the importance of understanding sex differences, and incorporating this knowledge in teaching our children.[5]

Same Behavior, Different reasons. Irving Sigel pointed out and gave some proof that even when boys and girls appear to behave similarly, each group does it for different reasons.[6] When one sex strives for power in a situation it may be to compensate for feelings of inadequacy, while the other sex strives for power because it is consistent with his sex role and/or biological drives. Sigel suggested his findings be remembered when future studies of sex differences are undertaken.

Sex Differences in Learning. Jean Grambs and Walter Waetjen, [7] wrote in *The National Elementary Principal* about the students' right of "Being Equally Different." They begin by mentioning that "the most unequal thing that happens in our schools is that unequals are treated as equals." They are referring to sex differences:

> ... our schools are sex-neutral institutions operating on the assumption that all persons are alike with respect to the ways in which they learn and achieve. We wish to make crystal clear our position that *it makes a significant difference whether the person we are teaching is a boy pupil or a girl pupil and that instructional provisions should be made accordingly*." (pp. 59-60)

Note: See chapter 6 for more on this subject.

50 Percent Hiring Demand Issue

Radical feminists demand that employers hire at least 50% women. They especially demand this in all good-paying positions, or status positions. They almost never ask for 50% of men's poor-paying, or poor status, or very boring jobs. They furthermore are not trying to balance female occupations to the 50% level for both sexes. To give women 50% of all job openings is not fair to men. See Chapter 2 for more details.

Wage Difference Issue

Most radical feminists make claims about sex discrimination in wages. They tell us that women working full time earn on the average only 77% of what men earn in the USA.[10] They tell us this discrepancy in wages is because of sexual discrimination. When some aspects such as overtime pay and union affiliation are taken into consideration, women still earn only about 91%[112] of what men earn. Because of male's biased value system women are discriminated against in the work place. Less so today than in the past. So the movement of feminism has done some good in this area.

Pregnancy and Work Leave Issue

Radical women liberators were making demands for "maternity lib" rights in the early 1970s. From a Wall Street Journal 1972 article we read:

---

[112] Washington Post, June 5, 2012 — http://www.washingtonpost.com/blogs/wonkblog/post/women-earn-91-cents-for-every-dollar-men-earn--if-you-control-for-life-choices/2012/06/04/gJQAqrHkEV_blog.html

8. Summary of Evidence and Issues

"They want to choose, without interference from the boss, when they'll quit work to have a baby and when they'll return. They want disability pay while unable to work. They seek more generous employer contributions for insurance covering their doctor and hospital bills. And they don't want an argument about getting their old job back when they return to work."[11]

The Equal Employment Opportunity Commission compared the demand for maternity compensation with men's financial compensation from the state or their employers when they are injured or sick and can't work.[12] But paid maternity leave would be in addition to state compensation for injury or unemployment. This paid maternity leave would be discriminatory against men unless men were also given "maternity" leave. Since 1972 there have been laws passed to give women unpaid leave and some corporations also give more generous maternity leaves.

There are pluses and drawbacks to this.

The plus is that today, most women work, and if you give them paid maternity leave, they would at least have six or so months to be with their babies and thus give their babies the love and attention they need. There would be less mother deprived babies. (see Chapter 6)

But policies would and have increased the costs of products and taxes (if maternal leave was paid by the state). This in turn could force other mothers out into the labor market who previously stayed home to care for their family. It may be a no win situation.

Divorce, Child Support, and Alimony Issue

In our society divorce is almost encouraged in many cases. But it would be better for our society to de-emphasize divorce. That is, our society should encourage communication and love between couples, and discourage divorce. Courses should be initiated in high school on sex differences. In this way, each sex can better understand the actions and desires of the other sex. Little is now done to prepare our children for marriage. This is reckless behavior on our part, for marriage has been and will be an essential part of our civilization. When people are not prepared for marriage, and not encouraged to strive within marriage for success, then divorces become more prevalent. And the problem with divorce is that no one wins. The children suffer. The husband suffers. The wife suffers. The society suffers.

Families of average and below average income cannot survive financially after divorce without aid from other tax payers through such means as child care centers and welfare support. Divorce not only affects the people obtaining a divorce, but it affects society on the whole. When one of our neighbors gets divorced, directly and indirectly, our taxes increase. When our society is hit with a high divorce rate, we are forced to work harder in order to pay the higher taxes. Our wives are forced to enter the job market, thus, leaving our homes with inferior processed foods, inferiorly reared children, inferior home atmospheres, and inferior family interaction. Therefore, divorces should be actively discouraged; harmony in marriage should be encouraged.

When Divorce Comes. But if and when divorce comes to a couple, there is the matter of child support and alimony. The average employed man can't afford to support an ex-wife, children and himself in separate households. It is unfair to ask society to support them. It is unfair to ask the husband to support his ex-wife, if she gives nothing in return to him. It is unfair for the children to be reared in such a situation. It is unfair to ask the wife to support the children alone, or to support her ex-mate. But it is fair in divorce to ask the wife to support herself, the husband to support himself, and both to support the children. Both ex-mates should give time, money, and love to their children. One or the other should not have the sole burden, but both should be asked by society to support their children and themselves. Many radical feminists think it is man's task to support the children. This is unfair discrimination. Both should support the children (not necessarily equally) as long as women are given equitable opportunity in the job market and wages are equitable.

Abortion Issue

Radical feminists believe that each and every woman has the "right" to abortion. They imply and even openly say that every woman should have the sole right to decide on an abortion even without the consent of her husband. And the Supreme Court of the United States, on July 1, 1976, not only ruled that a woman does not have to get her husband's consent for an abortion, but that teenage girls do not have to get their parents permission for an abortion. Notice the discriminatory factor of this as related by M.J. Sobran, in his article, "of Ms. and Men":

8. Summary of Evidence and Issues

> "John and Jane Doe, say, marry; Jane gets pregnant; John wants the child; Jane doesn't, and exercises her unqualified prerogative of getting it aborted; and that's that. Now take the opposite case: Jane is pregnant: John doesn't want the kid; Jane does, and refuses to abort it; they divorce; John has to support the kid he never wanted."[15]

Besides this discriminatory factor, one must remember that if our parents aborted us we would not be here. It is much better to use birth control to control family size. Birth control should be emphasized, not abortion. Abortion is a destructive act; birth control is a preventive act.

Sexual Freedom Issue

Some women liberators call for sexual freedom.

They say the only reason women are not sexually free now is because of their past puritanical rearing by non-liberated parents. We believe such radical feminists are projecting a level of ignorance of women's sexuality.

Dr. Margaret Deanesly, internist at Cowell Student Health Center (Stanford), has concluded that today's changed sexual morals and easy access to birth control are no panacea to young women:

> "If I were convinced that the new morality meant stability and serenity for a large population of girls, I would be all for it. But judging from the people I see, I'm not convinced."[16]

Sexual freedom may sound good to some, but few achieve it without psychological and physical damage.

Dress Issue

Although at present in the U.S.A. the wearing of pants among women is common, it was not so in the 1800s or early in the 1900s. Looking back we can see why the idea of pants wearing by women developed. One of the reasons has to do with feminist dogma. From letters exchanged in 1855 between Gerrit Smith and Elizabeth Cady Stanton, we can get an idea of this dogma: according to Smith, women's dress "imprisons and cripples them;" according to Stanton, "it seems that if she would enjoy entire freedom, she should dress just like man."[17] So these radical feminists believed that women's dresses limit women somehow, and to enjoy freedom women must be able to dress like men. Yet, a lot of men in the Middle East and elsewhere also wear "dresses" (the thawb, suriyah, etc.). Are they imprisoned?

But some differences in clothing styles by the sexes is needed for at least one important reason: the gender identity of children. In order to have a psychosexual healthy atmosphere, each sex must identify with its own.[18] It is psychologically better for boys to identify with men, and girls to identify with women. If everyone went around nude, there would not be any problem here, for a nude man looks obviously different from a nude woman.

Sexual Maturity

This is the true meaning of being sexually mature: understanding the other sex's differences, accepting them as real, giving love as the other needs and wants. The two should learn to harmonize, combine, and become one as much as possible without radically mixing real sex roles that are based on biology. Let's recognize that both sexes are self-centered in some ways, yet each sex should learn from the other, and harmonize with the other. Such maturity will lead to healthier relations between the sexes.

References for Chapter 8

[1: ch 8] p. 453 in <u>4</u> of <u>Bibliography</u> list
[2: ch 8] <u>86</u>
[3: ch 8] <u>86</u>
[4: ch 8] <u>23</u>
[5: ch 8] <u>61</u>; <u>162</u>; <u>69</u>
[6: ch 8] <u>162</u>
[7: ch 8] <u>69</u>
[8: ch 8] <u>108</u>
[9: ch 8] <u>108</u>
[10: ch 8] see latest Almanac
[11: ch 8] <u>87</u>
[12: ch 8] <u>158</u>; <u>195</u>
[13: ch 8] <u>175</u>
[14: ch 8] <u>175</u>
[15: ch 8] p. 581 in <u>170</u>
[16: ch 8] <u>176</u>
[17: ch 8] <u>95</u>
[18: ch 8] <u>120</u>

9. The Biblical Perspective

What does the Bible say about males and females and sex?

If you have read the first 8 chapters of this book you will see the scientific studies on sex and sex differences. There are not 100s of sexes/genders — only two. This evidence and the evidence of the sex differences manifested throughout the history of mankind makes the the following Biblical statement agreeable to functional knowledge:

In the Holy Bible, Genesis 5:2 it says:

"Male and Female He Created them"

Of course, some will disagree who or what God was/is/will-be, but that is for another book.[113] For anyone to disagree is put them on the side of the blatantly blind, physically and spiritually.

Male and Female are two who became one:

> Gen 26 And Gods[114] said, Let us make man in our image, after our likeness: and let them have dominion over the fish of the sea, and over the fowl of the air, and over the animal, and over all the earth, and over every creeping thing that creeps, upon the earth 1:27 So Gods[115] creates[116] the man[117] in his own image, in the image of Gods[114] he created[118] him; male and female He created[117] them.

After you understand the footnotes, read this:

After God formed Adam out of the dust of the ground he breathed into his nostrils the breath of life and thus Adam became a living

---

[113] *My God is the Becoming-One*, ISBN: 9781619180543

[114] Heb. Plural word (*elohim*) that mean godS, either two or more, but points to Ex 25:17-22 and the **two** cherubs in the Holy of Holies. See our "Image of God Paper"

[115] the Hebrew plural noun *elohim*, in context meaning two (not three) spiritual beings as depicted by the two cherubs in the Holy of Holies.

[116] incomplete Hebrew verb is a verb of incomplete or imperfect action

[117] Adam

[118] complete Hebrew verb, a verb of completed or perfect action

MALE AND FEMALE HE CREATED THEM

being, a soul (Gen 2:7). He said it was not good that Adam to be alone (Gen 2:18). God then formed out of the ground the various animals, but among them God could not find a corresponding[119] helper for Adam among the animals (Gen 2:19-20). So God put Adam in a deep sleep and in his sleep He took a piece from one of Adam's sides and made that side into a woman and brought her to Adam (Gen 2:21-23). And they became one flesh (Gen 2:24). When God created man, He made him in the likeness of God, Male and female he created them and called their name Adam (Gen 5:1-2). Notice he called both of them Adam, two became one flesh. This means both are in the image of the Gods. The image of the God Power was in the likeness of the Gods (plural): there were two, male and female, and He called both of them Adam, and both became one.

When we actually translate the literal meaning of the Hebrew words, we see that the powerful God of the OT was plural, more than one, and that the power was represented by **two** cherubs (two angels) in the Holy of Holies. The two cherubs were made from hammered work of pure gold and were joined as one piece to the pure gold mercy seat/lid of the Ark of the Covenant (Ex 25:17-22).

The Bible clearly depicts throughout that there are only **two** sexes: male and female.

1. Marriage in the Bible is only between a male and female as are the general laws and behavior of mankind throughout history.

2. Bible lists OT laws against dressing in the opposite sexes' clothing, thus projecting only two sexes.

3. God telling Adam and Eve to be fruitful and multiply (Gen 1:28) which can only happen between a male and female.

4. God said he created them as corresponding or opposite to each other, not the same (Heb. Gen 2:18, 20).

Males and Females He Created Them: All studies, laws, teachings that state otherwise are dysfunctional, illogical, non-scientific and spiritually lawless.

---

[119] Hebrew word that also means "opposite" "in front of"

BIBLIOGRAPHY

Note: for this edition, please see the 100+ footnotes in the text and the many references within brackets and parentheses in the text that refer to the books in this bibliography and newer books and studies pertaining to sex/gender differences that are not listed in this bibliography.

1 ABBOTT, SIDNEY and BARBARA LOVE. Is Women's Liberation a Lesbian Plot? In (67)

2 AINSWORTH, MARY D. The Effects of Maternal Deprivation: A Review of Findings and Controversy in the Context of Research Strategy. In (210)

3 Almost Half the Wives in U.S. Have Jobs Now. *San Jose Mercury-News*, Sunday May 23, 1976, p. 2.

4 ANASTASI, ANNE. *Differential Psychology*. (3rd ed.) MacMillan: New York, 1965.

5 ANDERSON, C.C. A Developmental Study of Dogmatism During Adolescence with Reference to Sex Differences. *Journal of Abnormal and Social Psychology*, 1962, Vol. 65, No. 2, pp. 132-135.

6 ANDREW, R.J. and L. ROGERS. Testosterone — Effects on Search Behaviour and Persistence. *Nature*, 1972, Vol. 237, No. 5354, pp. 343-345.

7 ARCHER, JOHN. Sex Differences in Emotional Behaviour: A Reply to Gray and Fuffery. *Acta Psychologica*, 1971, Vol. 35, pp. 415-429.

8 BADCHUK, NICHOLAS and ALAN P. BATES. Professor or Producer: The Two Faces of Academic Man. *Social Forces*, 1962, Vol. 40, pp. 341-348.

9 BARDWICK, JUDITH M. *Psychology of Women: A Study of Bio-Culture Conflicts*. Harper & Row: New York, 1971.

10 BARDWICK, JUDITH M. (ed.) *Readings on the Psychology of Women*. Harper & Row: New York, 1972.

11 BARDWICK, JUDITH M. and ELIZABETH DOUVAN. Ambivalence: The Socialization of Women. In (10).

12 BARDWICK, JUDITH M. Her Body, the Battleground. *Psychology Today*, Feb. 1972, Vol. 5, No. 9, pp. 50ff.

13 BARLOW, DAVID H., E. JOYCE REYNOLDS, and STEWART AGRAS. Gender Identity Change in a Transexual. *Archives of General Psychiatry*, 1973, Vol. 28, pp. 569-576.

14 BARRY III, HERBERT, MARGARET K. BACON, and IRVIN L. CHILD. A Cross-Cultural Survey of Some Sex Differences in Socialization. *Journal of Abnormal and Social Psychology*, 1957, Vol. 55, No. 3, pp. 327-332.

15 BARTELL, G.D. Group Sex Among the Med-Americans. *Journal of Sex Research*, 1970, Vol. 6, pp. 113-130.

16 BARUCH, GRACE K. Maternal Influences Upon College Women's Attitudes Toward Women and Work. *Developmental Psychology*, 1972, Vol. 6, No. 1, pp. 32-37.

17 BAYLEY, NANCY. Individual Patterns of Development. *Child Development*, 1956, Vol. 27, No. 1, pp. 45-74.

18 BEACH, FRANK A. A Review of Physiological and Psychological Studies of Sexual Behavior in Mammals. *Physiological Reviews*, 1947, Vol. 47, pp. 240-307.

19 BENEDICT, RUTH. Continuities and Discontinuities in Cultural Conditioning. *Psychiatry*, 1938, Vol. 1, pp. 161-167.

20 BENEDICT, RUTH. *Patterns of Culture.* New American Library: New York, 1960. (First published in 1934 by Houghton Mifflin Co.)

21 BENNETT, GEORGE K. and RUTH M. CRUIKSHANK. Sex differences in the Understanding of Mechanical Problems. *Journal of Applied Psychology*, 1942, Vol. 26, No. 1, pp. 121-127.

22 BEN YOSEF, AVRAHAM C. *The Purest Democracy in the World.* Herzl Press & Thomas Yoseloff: New York, 1963.

23 BERG, ALAN. The Economics of Breast-Feeding. *Saturday Review/The Sciences*, May 1973, Vol. 1, No. 4, pp. 29ff.

24 BERG, IAN, HAROLD H. NIXON and ROBERT MACMAHON. Change of Assigned Sex at Puberty. *The Lancet*, Dec. 7, 1963, pp. 1216-1217.

25 BERNAL, JUDITH and P.M. RICHARDS. Effect of Bottle and Breast Feeding on Infant Development. *Journal Psychosomatic Research*, 1970, Vol. 14, pp. 247-252.

26 BOWLBY, JOHN. *Maternal Care and Mental Health.* World Health Organization; Geneva, 1952.

27 BOWLBY, JOHN. *Separation: Anxiety and Anger.* (vol. 2 of *Attachment and Loss*) Basic Books: New York, 1973.

28 Breast Vs. Bottle. *Parade*, July 4, 1976, p. 20.

28a BRIZENDINE, LOUANN. *The Female Brain.* Broadway Books [Random House]: New York, 2006.

29 BROVERMAN, DONALD M. et al. Roles of Activation and Inhibition in Sex Differences in Cognitive Abilities. *Psychological Review*, 1968, Vol. 75, No. 1, pp. 23-50.

30 BRUN-GULBRANDSEN, SVERRE. Sex Roles and the Socialization Process. In **36**.

31 BUDNER, S. and JOHN MEYER. Women Professors. Unpublished manuscript cited in R.J. Somon, S.M. Clar, and K. Galway, The Woman Ph. D.: A Recent Profile, *Social Problems*, Fall 1967, Vol. 15, No. 2, pp. 221-236.

32 BURG, ALBERT. Visual Acuity as Measured by Dynamic and Static Tests. *Journal of Applied Psychology*, 1966, Vol. 50, No. 6, pp. 460-466.

33 CARMICHAEL, LEONARD (ed.). *Manual of Child Psychology* (2nd ed.), John Wiley: New York, 1954.

34 CHARTHAM, ROBERT. Masturbation Techniques — What Every Woman Should Know. *Forum*, Sept. 1972, Vol. 1, No. 12, pp. 14ff.

35 CHODOROW, NANCY. Being and Doing: A Cross-Cultural Examination of the Socialization of Males and Females. In (**67**).

36 DAHLSTROM, EDMUND (ed.). *The Changing Roles of Men and Women.* Beacon Press: Boston, 1971. [paperback]

37 DAHLSTROM, EDMUND. Analysis of the Debate on Sex roles. In (**36**), pp. 170-205.

38 D'ANDRADE, ROY G. Sex Differences and Cultural Institutions. In (**100**), pp. 173-203.

39 DAVID, M. and G. APPELL. A Study of Nursing Care and Nurse-Infant Interaction. In B.M. Foss (ed.), *Determinants of Infant Behaviour*, Metheun: London, 1961.

40 DAVID, M. and G. APPELL. Etude des Facteurs de Carence Affective Dans Une Pouponniere. *Psychiat. Enfant*, Vol. 4, Fasc. 2 (in press).

41 DAVIDSON, SARA. Foremothers. *Esquire*, July 1973, Vol. 80, No. 1, pp. 71ff.

42 De BEAUVOIR, SIMONE. *The Second Sex*. Bantam Books: New York, 1961. [First published in France in 1949]

43 De TOCQUEVILLE, ALEXIS. *Democracy in America*, Vol. 1l, Book 3, Chap. 12.

44 DEWHURST, C.J. and R.R. GORDON. Change in Sex. *The Lancet*, Dec. 7 1963, pp. 1213-1216.

45 DIAMOND, MILTON. A Critical Evaluation of the Ontogeny of Human Sexual Behavior. *Quarterly Review of Biology*, 1965, Vol. 40, pp. 147-175.

46 DIXON, MARLENE. The Rise of Women's Liberation. In (**10**).

47 DURDEN-SMITH, JO and deSIMONE, DIANE, *Sex and the Brain*, Warner Books:New York, 1984 (1983)

47a ECKERT, H.M. Visual-Motor Tasks at 3 and 4 Years of Age. *Percept. Mot. Skills*, 1970, Vol. 31, p. 560.

48 ECKERT, RUTH E. and JOHN E. STECKLEIN. *Job Motivations and Satisfactions of College-Teachers*, U.S. Dept. of Health, Education and Welfare, Office of Education, Cooperative Research Monograph No. 7, 1961.

49 ELIASBERG, ANN. Are you Harming Your Son Without Knowing it? *Family Circle*, April 1972, p. 44ff.

50 ELIOT, GEORGE. *Felix Holt, The Radical*. Penguin Books: Baltimore, Md., 1972. [First published in 1866]

51 EVERITT, B.F. and J. HERBERT. Adrenal Glands and Sexual Receptivity in Female Rhesus Monkeys. *Nature*, 1969, Vol. 222, pp. 1065-66.

52 FARBER, SEYMOUR M. and ROGER H.L. WILSON (eds.). *Man and Civilization: The Potential of Women*, Mcgraw-Hill: New York, 1963. [Paperback]

53 FIELD, MARK G. and DAVID E. ANDERSON. Family and Social Problems. In *Prospects for Soviet Society*, Allen Kassof (ed.), Praeger: New York, 1968, pp. 386-417.

54 FIELD, MARK G. and KARIN I. FLYNN. Worker, Mother, Housewife: Soviet Woman Today. In (**10**)

55 FOGARTY, M. et al. *Women in Top Jobs*. Allen & Unwin: London, 1971.

56 *FORUM: The International Journal of Human Relations*. [Conclusions drawn from reading *Forum* from October 1971 to 1973]

57 FRIEDAN, BETTY. Up From the Kitchen Floor. *The New York Times Magazine*, March 4, 1973, pp. 8ff.

58 GADPAILLE, WARREN J. Research into the Physiology of Maleness and Femaleness. *Archives of General Psychiatry*, 1972, Vol. 26, pp. 193-206.

59 GADPAILLE, WARREN J. Innate Masculine-Feminine Differences. *Medical Aspects of Human Sexuality*, 1973, Vol. 7, No. 2, pp. 140-156.

60 GANTT, W.H. Psychosexuality in Animals. In P.H. Hock and J. Zubin (eds.), *Psychosexual Development in Health and Disease*. Grune & Stratton: New York, 1949.

61 GARAI, JOSEF E. and AMRAM SCHEINFELD. Sex Differences in Mental and Behavioral Traits. *Genetic Psychology Monographs*, 1968, Vol. 77, 2nd half, pp. 169-299.

62 GARCIA, JOHN. IQ: The Conspiracy. *Psychology Today*, Sept. 1972, Vol. 6, No. 4, pp. 40ff.

63 GELLIFFE, D.B. and E.F.P. GELLIFFE (eds.). The Uniqueness of Human Milk. *American Journal of Clinical Nutrition*, 1971, Vol. 24, pp. 968- 1024.

64 GERSON, MENACHEM. Women in the Kibbutz. *American Journal of Orthopsychiatry*, 1971, Vol. 41, No. 4, pp. 566-573.

65 GOLDBERG, P. Misogyny and the College Girl. Paper presented at the meeting of the Midwestern Psychological Association, Boston, April 1967. Cited in (**16**).

66 GORDON, THOMAS P. and IRWIN S. BERSTEIN. Seasonal Variation in Sexual Behavior of All-male Rhesus Troops. *American Journal of Physical Anthropology*, 1973, Vol. 38, pp. 221-226.

67 GORNICK, VIVIAN and BARBARA K. MORAN (eds.). *Woman in Sexist Society: Studies in power and powerlessness*. Signet: New York, 1972. [Paperback]

68 GORNICK, VIVIAN. Consciousness. In (**10**)

69 GRAMBS, JEAN D. and WALTER B. WAETJEN. Being Equally Different: A New Right For Boys and Girls. *The National Elementary Principal*, 1966, Vol. 46, No. 2, pp. 59-67.

70 GRANT, ANNETTE. The New Feminist Literature: A Critique. *Sexual Behavior*, 1972, Vol. 2, No. 9, pp. 15-18.

71 GREER, GERMAINE. What Turns Women On? *Esquire*, July 1973, Vol. 80, No. 1, pp. 88ff.

72 GUETZKOW, H. An analysis of the Operation of Set in Problem-solving Behavior. *Journal of Genetic Psychology*, 1951, Vol. 45, pp. 219-244.

73 GUILLEMIN, ROGER and ROGER BURGUS. The Hormones of the Hypothalamus. *Scientific American*, 1972, Vol. 227, No. 5, pp. 24-33.

73a HAMBURG, DAVID A."Psychobiological Studies of Aggressive Behaviour," *Nature* Vol 230, March 5, 1971

74 HAMILTON, JAMES B. The Role of Testicular Secretions as Indicated by the Effects of Castration in Man and by Studies of Pathological Conditions and the Short Lifespan Associated with Maleness. In *Recent Progress in Hormone Research* (Vol. 111), Academic Press: New York, 1948.

75 HAMPSON, JOHN L. and JOAN G. HAMPSON. The Ontogenesis of Sexual Behavior in Man. In *Sex and Internal Secretions* (3rd ed.), W.C. Young (ed.), Williams and Wilkins: Baltimore, Md., 1961.

76 HAMPSON, JOHN L. Determinants of Psychosexual Orientation. In *Sex and Behavior*, Frank A. Beach (ed.), Wiley: New York 1965.

77 HARLOW, HARRY F. The Heterosexual Affectional System in Monkeys. *American Psychologist*, 1962, Vol. 17, pp. 1-9.

78 HARLOW, HARRY F. Sexual Behavior in the Rhesus Monkey. In *Sex and Behavior*, Frank A. Beach (ed.), Wiley: New York, 1965.

79 HELLBRONER, ROBERT L. *An Inquiry into the Human Prospect*, Norton: New York, 1974.

80 HEINIG, JEAN. Women's Lib: A Survey of American Attitudes. *Sexual Behavior*, March 1972, Vol. 2, No. 3, pp. 18-19.

81 HELD, VIRGINIA. Reasonable Progress and Self-Respect. *The Monist*, 1973, Vol, 57, No. 1, pp. 12-27.

82 HERBERT, J. Hormones and Reproductive Behavior in Rhesus and Talapoin Monkeys. *Journal of Reproduction and Fertility*, 1970 (Suppl. 11), pp. 119-140.

83 HILL, Jr., THOMAS E. Servility and Self-Respect. *The Monist*, 1973 Vol. 57, No. 1, pp. 87-104.

83a HOBSON, J.R. "Sex Differences in Primary Mental Abilities,' *Journal of Educational Research*, Vol 41, pp. 126-132, 1947

84 HOWE, IRVING. The Middle-Class Mind of Kate Millett. In (**10**).

85 HUBER, JOAN (ed.). *Changing Women in a Changing Society*. University of Chicago Press: Chicago, 1973.

86 HUTT, CORINNE. *Males and Females*. Penguin Books, Baltimore, Md., 1972. [Paperback]

87 HYTT, JAMES C. Expectant Mothers. *The Wall Street Journal*, Dec. 1 1972, p. 1, Col. 1.

88 JANEWAY, ELIZABETH. The Weak Are the Second Sex. *Atlanic*, Dec. 1973, Vol. 232, No. 6, pp. 91ff.

89 KANOVSKY, ELIYAHU. *The Economy of the Isreali Kibbutz*. Harvard U. Press: Cambridge, 1966.

90 KELLER, SUZANNE. The Future Status of Women in America. In U.S. Commission on Population Growth and the American Future. *Demographic and Social Aspects of Population Growth*, Vol. 1 of *Population and the American Future*, Charles F. Westoff and Rober Parke, Jr. (Editors), Vol. 1, Government Printing Office: Washington, D.C., 1972.

91 KEYS, A.B. et al. *The Biology of Human Starvation*. University of Minnesota P.: Minneapolis, 1950.

92 KILMARTIN, ANGELA. Thrush. *Forum*, July 1974, pp. 35-36.

93 KINSEY, A.C. et al. *Sexual Behavior in the Human Female*. Saunders: Philadelphia, 1953.

94 KOSTICK, MAX MARTIN. A Study of Transfer: Sex Differences in the Reasoning Process. *The Journal of Educational Psychology*, 1954, Vol. 45, No. 8, pp. 449-458.

95 KRADITOR, AILEEN S. (ed.). *Up From the Pedestal: Selected Writings in the History of American Feminism*. Quadrangle Books: Chicago, 1970. [Paperback]

96 LAKOFF, ROBIN. Language and Woman's Place. *Language in Society*, April 1973, Vol. 2, pp. 45-80.

97 LANGMEIER JOSEF. Personalities of Deprived Children In (**123**).

98 LEON, DAN. *The Kibbutz*. Pergamon: New York, 1969.

99 MACCOBY, ELEANOR F. Women's Intellect. In (**52**)

100 MACCOBY, ELEANOR E. (ed.). *The Development of Sex Differences.* Stanford U. Press: Stanford, Ca., 1966.

101 MACCOBY, ELEANOR E. Sex Differences in Intellectual Functioning. In (**100**).

102 MACCOBY, ELEANOR E. Sex in the Social Order. Book review in *Science*, Nov. 2, 1973, Vol. 182, pp. 469-471.

103 MACCOBY, ELEANOR E. and CAROL N. JACKLIN. *The Psychology of Sex Differences.* Stanford U Press: Stanford, Ca., 1974.

104 MACMEEKEN, A.M. *The Intelligence of a Representative Group of Scottish Children.* University of London Press: London, 1939.

105 MANNES, MARYA. The Problems of Creative Women. In (**52**).

106 MANNES, MARYA. The String. *Newsweek*, August 12 1974, p. 15.

107 MASTER, WILLIAM H. and VIRGINIA E. JOHNSON. *Human Sexual Response.* Little, Brown: Boston, 1966.

108 MCDERMID, NANCY. Sex Role Stereotyping Pervasive in School Texts. *ACLU News* (N. Calif.), special supplement on Women's Rights, March- April 1974.

109 MEAD, MARGARET. *Sex and Temperament.* Dell, Laurel Edition: New York, 1968. First published in 1935, William Morrow, Pub.

110 MEAD, MARGARET. *Male and Female: A Study of the Sexes in a Changing World.* Dell, Laurel Edition: New York, 1968. First published in 1949, William Morrow, pub.

111 MEAD, MARGARET. Cultural Determinants of Sexual Behavior. In W.C. Young (ed.), *Sex and Internal Secretions* (3rd. ed.), Williams & Wilkins: Baltimore, Md., 1961.

112 MISCHEL, WALTER. A Social-Learning View of Sex Differences in Behavior. In (**100**).

112a MISCHEL, WALTER. Sex-Typing and Socialization. In (**33**).

113 MOIR, ANNE and JESSEL, DAVID, *Brain Sex*, Laurel: New York, 1992 (1991)

114 MONEY, JOHN. Components of Eroticism in Man: The Hormones in Relation to Sexual Morphology and Sexual Desire. *Journal of Nervous and Mental Disease*, 1961, Vol. 132, pp. 239-248.

115 MONEY, JOHN. Sex Hormones and Other Variables in Human Eroticism. In W.C. Young (ed.), *Sex and Internal Secretions* (3rd. ed.), Williams & Wilkins: Baltimore, Md., 1961.

116 MONEY, JOHN. Developmental Differentiation of Feminity and Masculinity Compared. In (**52**).

117 MONEY, JOHN. Cytogenetic and Pychosexual Incogruities with a Note on Space-Form Blindness. *American Journal of Psychiatry*, 1963, Vol. 119, pp. 820-827.

118 MONEY, JOHN. Sexual Dimorphism and Homosexual Gender Identity. *Psychological Bulletin*, 1970, Vol. 74, No. 6, pp. 425-440.

119 MONEY, JOHN and ANKE A. EHRHARDT. Prenatal Hormonal Exposure: Possible Effects on Behavior in Man. In R.P. Michael (ed.), *Endocrinology and Human Behavior*. Oxford University Press: New York, 1968.

120 MONEY, JOHN, ANKE A. EHRHARDT. *Man & Woman, Boy & Girl.* The Johns Hopkins University Press: Baltimore, MD., 1972

121 MONEY, JOHN and ANKE A. EHRHARDT, and D.N. MASICA. Fetal Feminization Induced by Androgen Insensitivity in the Testicular Feminizing Syndrome: Effect on Marriage and Maternalism. *Johns Hopkins Medical Journal*, 1968, Vol. 123, pp. 160-167.

122 MONEY J., J.G. HAMPSON and J.L. HAMPSON. Hermaphroditism: Recommendations Concerning Assignment of Sex, Change of Sex, and Psychologic Management. *Bulletin of The Johns Hopkins Hospital*, 1955, Vol. 97, pp. 301-319.

123 MONKS, F.J., WILLARD W. HARTUP and JAN De WIT (eds.). *Determinants of Behavioral Development*. Academic Press: New York, 1972.

124 MOSS, HOWARD A. Sex, Age, and State as Determinants of Mother-Infant Interaction. Merrill-*Palmer Quarterly*, 1967, Vol. 13, No. 1, pp. 19-36.

125 MOSTELLER, FREDRICK and DANEL P. MOYNIHAN. *On Equality of Educational Opportunity*. Vintage Books: New York, 1972.

126 MOTHERSILL, MARY. Notes on Feminism. *The Monist*, 1973, Vol. 57, No, 1, pp. 105-114.

127 MOYLAN, PETE. Age-Old Prejudices Still Face Women In Business. *San Jose Mercury*, April 22 1974, p. 18.

128 NADEL, S.F. The Typological Approach to Culture. *Character and Personality*, Vol. 5, p. 275.

129 NASH, JOHN. *Developmental Psychology: A Psychobiological Approach* Prentice-Hall: Engelwood Cliffs, N.J., 1970.

130 National Organization for Women. Statement of Purpose (1964). In (**95**).

131 NEWTON, NILES. Battle Between Breast & Bottle. *Psychology Today*, July 1972, Vol. 6, No. 2, pp. 68ff.

132 OETZEL, ROBERTA M. Annotated Bibliography. In (**100**).

133 OETZEL, ROBERTA M. Classified Summary of Research in Sex Differences. In (**100**).

134 O'NEILL, G.C. and N. O'NEILL. Patterns in Group Sexual Activity. *Journal of Sex Research*, 1970, Vol. 6, pp. 101-112.

135 OPPENHEIMER, V.K. Demographic Influence on Female Employment and the Status of Women. In (**85**).

136 OUNSTED, C. and D.C. TAYLOR (eds.). Gender *Differences*: *Their Ontogeny and Significance*. Williams & Wilkins: Baltimore, Md., 1972.

137 OVERSTREET, EDMUND W. Biological Make-up of Woman. In (**52**).

138 PARSONS, TALCOTT and ROBERT F. BALES. *Family*, *Socialization and Interaction Process*. The Free Press: Glencoe, Ill., 1955.

139 PEER, ELIZABETH. Dirty Words. In "Ideas," *Newsweek*, Feb. 4 1974, p. 78.

140 PERSKY, H., K.D. SMITH, and G.K. BASU. Relation of Psychologic Measures of Aggression and Hostility to Testostere Production in Man. *Psychosomatic Medicine*, 1971, Vol. 33, pp. 265-227.

141 PIDDINGTON, R. *An Introduction to Social Anthropology* (Vol. 1). Oliver & Boyd: Edinburgh, 1950.

142 PIDDINGTON, R. *An Introduction to Social Anthropology* (Vol. 2). Oliver & Boyd: Edinburgh, 1957.

142a REISMAN, JUDITH A. and EICHEL, EDWARD W., *Kinsey, Sex and Fraud: The Indoctrination Of A People*, Lochinvar-Huntington House Pub: Lafayette, Louisiana, 1990

143 RHEINGOLD, H.L. The Modification of Social Responsiveness in Institutional Babies. *Monogr. Soc. Res. Child Dev.*, 1956, Vol. 21, No. 63.

144 RIDLEY, JEANNE CLARE. On the Consequences of Demographic Change for the Roles and Status of Women. In *Demographic and Social Aspects of Population Growth*, Charles F. Westoff and Robert Parke, Jr. (eds.), Vol 1, Government Printing Office: Washington, D.C., 1972.

145 RODALE, ROBERT. Breast-Fed Babies Have an Advantage. *San Jose Mercury-News*, Sunday June 30 1974, p. 05.

146 ROSE, R.H., J.W. HMLADAY, and I.S. BERNSTEIN. Plasma Testosorone Dominance Rank and Aggressive Behavior in Male Rhesus Monkeys. *Nature*, 1971, Vol. 231, pp. 366-368.

147 ROSENZWEIG, MARK R., EDWARD L. BENNETT, and MARIAN CLEEVES DIAMOND. Brain Changes in Response to Experience. *Scientific American*, 1972, Vol. 226, No. 2, pp. 22-29.

148 ROSS, SUSAN C. *The Rights of Women: The Basic ACLU Guide to a Woman's Rights*. Avon Books: New York, 1973.

149 RUCH, FLOYD L. (ed.). *Psychology and Life*. Scott, Foresman & Co: Glenview, Ill., 1967.

150 RUTTER, MICHAEL. *Maternal Deprivation Reassessed*. Penguin Books: Baltimore, Md., 1972.

151 SCHEINFELD, AMRAM. *Women and Men*. Harcourt, Brace and Co.: New York, 1943.

152 SCHEINFELD, AMRAM. *Your Heredity and Environment*. Lippincott: Philadelphia, Pa., 1965.

153 SCHMIDT, G. and V. SIGUSCH. Sex Differences in Responses to Psychosexual Stimulation by Films and Slides. *Journal of Sex Research*, 1970, Vol. 6, pp. 268-283.

154 ACHMIDT, HARALD EDWIN. Spatial Creativity in Second Year Architecture Students. *Perceptual and Motor Skills*, 1970, Vol. 31, pp. 561-562.

155 SCHNEIR, MIRIAN. *Feminism: The Essential Historical Writings*. Random House: New York, 1972.

156 SEARS, ROBERT R. Development of Gender Role. In *Sex and Behavior*, Frank A. Beach (ed.), Wiley: New York, 1965.

157 SELIGMAN, MARTIN E.P. and JOANNE L. HAGER (eds.). *Bimlogical Boundaries of Learning*, Appleton-Century-Crofts: New York, 1972.

158 Sex Bias: Punishing Pregnancy. *Civil Liberties*, March 1974, No. 301. [ACLU national news]

159 SHAPIRO, ANDREW LL., *We're Number One*, Vintage Books, New York, 1992

160 SHEPHER, J. Self-Imposed Incest Avoidance and Exogamy in Second Generation Kibbutz Adults. Unpublished Doctoral Dissertation. Rutgers U.8 Brunswick, N.J., 1971. Cited in (**120**).

161 SHEPHER, J. Male Selection Among Second Generation Kibbutz Adolescents and Adults: Incest Avoidance and Negative Imprinting. *Archives of Sexual Behavior*, 1971, Vol. 1, pp. 293-307.

162 SIGEL, IRVING E. Rationale for Separate Analyses of Male and Female Samples on Cognitive Tasks. *The Psychological Record*, 1965, Vol. 15, pp. 369-376.

163 SIGUSCH, V. et al. Psychosexual Stimulation: Sex Differences. *Journal of Sex Research*, 1970, Vol. 6, pp. 10-24.

164 SILVERMAN, ANNA and ARNOLD. The Wrong Reasons for Motherhood. *Sexual Behavior*, May 1972, Vol. 2, No. 5, pp. 58-64.

165 SINGER, JUDITH E., MILTON WESTPHAL and KENNETH R. NISWANDER. Sex Differences in the Incidence of Neonatal Abnormalities and Abnormal Performance in Early Childhood. *Child Development*, 1968, Vol. 39, pp. 103-122.

166 SIMON, RITA JAMES, SHIRLEY MERRITT CLARK and KATHLEEN GALWAY. The Woman Ph. D.: A Recent Profile. *Social Problems*, 1967, Vol. 15, No. 2. Cited from 10, pp. 83-92.

167 SMITH, GERRIT. Correspondence Between Gerrit Smith and Elizabeth Cady Staton (1855). In 95.

168 SMITH, HEDRICK. *The Russians*. Quadrangle/The New York Times Book Co.: New York, 1976.

169 SMITH, MACFARLANE. *Spatial Ability*. Robert R. Knapp: San Diego, Ca., 1964.

170 SOBRAN, Jr., M.J. Of Ms. and Men. *National Review*, May 24, 1974, pp. 579-581.

171 Social Security: Promising Too Much to Too Many? *U.S. News & World Report*, July 15 1974, Vol. 77, No. 3, pp. 26-30.

172 Soviet Sexist Attitude Horrifies NOW President. *San Jose Mercury*, Thur. Oct. 16 1975, p. 57.

173 SPIRO, M.E. *Kibbutz: Venture in Utopia*. Harvard University Press: Cambridge, 1956. Quoted from (38).

174 SPIVAK, JONATHAN. Social Security Financially Hurt by Lag in Births. *The Wall Street Journal*, June 4 1974, p. 2.

175 Sportswomanlike Conduct. *Newsweek*, June 3 1974, pp. 50ff.

176 Stanford Coeds and Sex. *San Jose News*, April 14 1972, pp. 1ff.

177 STASSINOPOULOS, ARIANNA. *The Female Woman*. Random House: New York, 1973.

178 The Status of Women in Sweden: Report to the United Nations 1968. In (36).

178a STEEN, EDWIN B. and PRICE JAMES H. *Human Sex and Sexuality*, 2nd Edition, Dover Pub: New York, 1988

179 STEIN, SARA and CARTER SMITH. Return of Mom. *Saturday Review/Education*, April 1973, Vol. 1, No. 3, pp. 37ff.

180 STODDARD, GEORGE D. *The Meaning of Intelligence*. Macmillan: New York, 1943.

181 STONE, LUCY. Lucy Stone, Speech (1855). In (95).

182 SWEENEY, E.J. Sex Differences in Problem Solving. Technical Report No. 1, Dept. of Psychology, Stanford University, Stanford, Ca. 1953.

183 TAVRIS, CAROL. Women in China: The Speak Bitterness Revolution. *Psychology Today*, May 1974, Vol. 7, No. 12, pp. 43ff.

184 TERMAN, LEWIS M. and CATHARINE COX MILES. *Sex and Personality: Studies in Masculinity and Femininity*. Mcgraw-Hill: New York, 1936.

185 TERMAN, LEWIS M. and LEONA E. TYLER. Psychological Sex Differences. In (33).

186 Test-Tube Babies? *Newsweek*, July 29 1974, p. 70.

187 "The Forum." *Forum*, July 1974, p. 70.

188 TILLER, PER OLAV. Parental Role Division and the Child's Personality Development. In (36).

189 TIMASHEFF, NICHOLAS S. The Attempt to Abolish the Family in Russia. In *A Modern Introduction to the Family*, Norman W. Bell and Ezra F. Vogel (eds.), Free Press: Glencoe, Ill., 1960.

190 TURNBULL, COLIN M. *The Mountain People*. Simon and Schuster: New York, 1972.

191 TYLER, LEONA E. *The Psychology of Human Differences* (3rd. ed.). Appleton-Century-Crofts: New York, 1965.

192 ULMER, MELVILLE J. *Economics: Theory and Practice* (2nd ed.). Houghton Mifflin: Boston, 1965.

193 U.S. Bureau of the Census. *Statistical Abstract of the United States* (93rd ed.). Government Printing Office: Washington, D.C., 1972.

194 U.S. Bureau of the Census. *Statistical Abstract of the United States* (95th ed.). Government Printing Office: Washington, D.C., 1974.

195 U.S. Equal Employment Opportunity Commission. Guidelines on Discrimination because of Sex. In *Affirmative Action and Equal Employment* (Vol. 2). Government Printing Office: Washington, D.C., 1974.

196 U.S. House of Representatives. Committee on the Judiciary. *Equal Rights for Men and Women*. Government Printing Office: Washington, D.C., 1971.

197 U.S. Treasury Department. *Statistics of Income — 1962, Personal Wealth Estimated from Estate Tax Returns*. Government Printing Office: Washington, D.C., 1967.

198 WATSON, JOHN S. Operant Conditioning of Visual Fixation in Infants Under Visual and Auditory Reinforcement. *Developmental Psychology*, 1969, Vol. 1, No. 5, pp. 508-516.

199 WATTS, WILLIAM and FLOYD A. FREE (eds.). *State of the Nation*. Patomac Assoc., 1973.

200 WECHSLER, DAVID. *The Measurement of Adult Intelligence* (3rd ed.). Williams & Wilkins: Baltimore, Md. 1944.

201 WEISSTEIN, NAOMI. Psychology Constructs the Female. In (67).

202 WHALEN, RICHARD E. Hormones and Behavior. In *Hormones and Behavior*, R.E. Whalen (ed.), D. Van Nostrand: New York, 1967. [Paperback]

203 "What's News." *The Wall Street Journal*, July 10 1974, p. 1, Col. 3.

204 WHITE, SUZANNE. Verbal Intimacy. *Forum*, March 1974, pp. 26-31.

205 WITKIN, H.A. et al. *Personality Through Perception.* Harper & Brothers: New York, 1954.

206 WONG, WILLIAM. Axing Alimony. *The Wall Street Journal*, May 29, 1974, p. 1.

207 WOODWARD, KENNETH L. Do Children Need Sex Roles? In "Ideas," *Newsweek*, June 10 1974, pp. 79ff.

208 WORK, MURRAY S., NEAL GROSSEN, and HILLIARD ROGERS. Role of Habit and Androgen Level in Food-Seeking Dominance Among Rats. *Journal of Comparative and Physiological Psychology*, 1969, Vol. 69, No. 4, pp. 601-607.

209 WORK, MURRAY S. and HILLIARD ROGERS. Effect of Estrogen Level on Food-Seeking Dominance Among Male Rate. *Journal of Comparative and Physiological Psychology*, 1972, Vol. 79, No. 3, pp. 414-418.

210 World Health Organization. *Deprivation of Maternal Care: A Reassessment of its Effects.* Public Health Papers: Geneva, 1962.

211 WRIGHT, JOHN W., Editor, *The Universal Almanac 1992*, Andrews and McMeel: New York, 1991

212 YALOM, IRVIN D., RICHARD GREEN and NORMAN FISK. Prenatal Exposure to Female Hormones. *Archives of General Psychiatry*, 1973, Vol. 28, pp. 554-562.

213 YARROW, LEON J. Enrichment and Deprivation: Towards a Conceptual and Empirical Differentiation of the Early Environment. In (**123**).

214 Zero Population Growth Can Be Economic Jolt. *San Jose Mercury-News*, Sunday April 14 1974, p. 3f.

215 Personal observation.

216 Personal observation and secondary source: Shearer, Lloyd. In Other Words. *Parade*, May 7 1972, p. 11.

Author
Walter R. Dolen

I gratefully thank my wife, Shirley Clare, for editing the first edition of this book, and I thank my daughter for editing the second edition of this book.

About the Author

Walter Dolen is an author/editor of several books, using the scient method[120]including: *My God is the BeComingOne: God Papers*; *New Mind Papers*; *6000 Years of Mankind*; *BeComing-One Bible*; *Harmony of the Gospels*; *Harmony of the Good News*; *Male & Female*; *Prophecy Papers*; *Einstein Light Time Relativity;* etc. These books were researched and written between 1969 and 2025. Walter has worked with his hands (carpenter/builder), with his mind (publisher/ writer/ building designer) and with his soul (President of the Becoming-One Church).

For more information about the author see his web site:
www.walterdolen.com or www.walterdolen.ws

---

[120] (1) Perceive a problem; (2) examine and analyze all the available evidence; (3) examine and imagine different hypotheses in attempt to solve the problem in a logical manner; (4) form a theory that answers the problem; (5) test the theory; (6) always have an open mind for better theories or answers to the problem; (7) change the theory if new evidence is inconsistent to your prior theory.

www.ingramcontent.com/pod-product-compliance
Lightning Source LLC
Chambersburg PA
CBHW070057080526
44586CB00013B/1097